The Org

Rea

I'm an online organic chemistry tutor. Over the past two years I've spent over two thousand hours coaching students in organic chemistry courses. One of the most consistent complaints my students express to me is what a nightmare it is to keep track of the vast number of different reagents in their organic chemistry course. I found myself answering the same questions again and again: "What is DIBAL?", "What does DMSO do?", "What reagents can I use to go from an alcohol to a carboxylic acid?". While textbooks indeed do contain this information, the important contents can be scattered throughout a 1000+ page tome. Furthermore, online resources like Wikipedia are often not aimed at the precise needs of the student studying introductory organic chemistry.

I thought it would be useful to take all the reagents that students encounter in a typical 2-semester organic chemistry course and compile them into a big document. Hundreds of hours of work later, the result is before you: "The Organic Chemistry Reagent Guide".

This document is divided into three parts:

Part 1: Quick Index of Reagents. All the key reagents and solvents of organic chemistry on one page.

Part 2: Reagent profiles. Each reagent (>80 in all) has its own section detailing the different reactions it performs, as well as the mechanism for each reaction (where applicable).

Part 3: Useful tables. This section has pages on common abbreviations, functional groups, common acids and bases, oxidizing and reducing agents, organometallics, reagents for making alkyl and acid halides, reagents that transform aromatic rings, types of arrows, and solvents.

I would like to thank everyone who has helped with proofreading and trouble-shooting, in particular Dr. Christian Drouin whose contributions were immensely valuable. I would also like to thank Dr. Adam Azman, Shane Breazeale, Dr. Tim Cernak, Tiffany Chen, Jon Constan, Mike Evans, Mike Harbus, Dr. Jeff Manthorpe, for helpful suggestions, along with countless readers who reported small errors and typos in the first edition.

The primary references used for this text are “Organic Chemistry” by Maitland Jones Jr. (2nd edition) and “March’s Advanced Organic Chemistry” (5th edition.

This work is continually evolving. Although considerable effort has been expended to make this as thorough as possible, no doubt you will encounter reagents in your course that are not covered here. Please feel free to suggest reagents that can be included in future editions. Furthermore you may also find some conflicts between the material in this Guide and that in your course. Where conflicts arise, your instructor is the final authority.

Any errors in this document are my own; I encourage you to alert me of corrections by email at james@masterorganicchemistry.com

Above all, else: I hope this Guide is useful to you!

And if you have any suggestions or find mistakes, please leave feedback

Sincerely, James A. Ashenhurst, Ph.D.
Founder, MasterOrganicChemistry.com
James@masterorganicchemistry.com
Twitter: @jamesashchem

Index

Reagents and Solvents

Index

End Notes

$AgNO_3$
Silver Nitrate

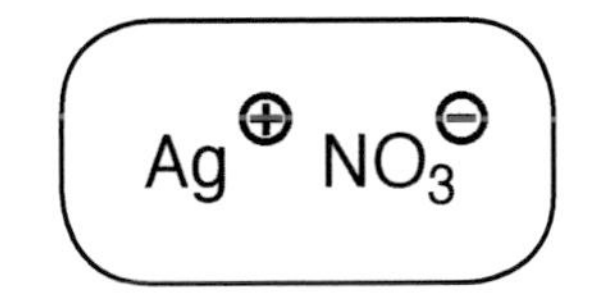

What it's used for: Silver nitrate will react with alkyl halides to form silver halides and the corresponding carbocation. When a nucleophilic solvent such as water or an alcohol is used, this can result in an S_N1 reaction. It can also react in the Tollens reaction to give carboxylic acids from aldehydes.

Similar to: $AgBF_4$

Example 1: Substitution (S_N1) conversion of alkyl halides to alcohols

Br → ($AgNO_3$, H_2O) → OH + HNO_3 + AgBr

Example 2: Substitution (S_N1) conversion of alkyl halides to ethers

Cl → ($AgNO_3$, MeOH) → OMe + HNO_3 + AgCl

Example 3: Tollens oxidation - conversion of aldehydes to carboxylica acids

O, H → ($AgNO_3$, NH_3, H_2O) → O, OH

How it works: ***S_N1 Reaction of alkyl halides***

Silver nitrate, $AgNO_3$, has good solubility in aqueous solution, but AgBr, AgCl, and AgI do not. Ag^+ coordinates to the halide, which then leaves, forming a carbocation. The carbocation is then trapped by solvent (like H_2O)

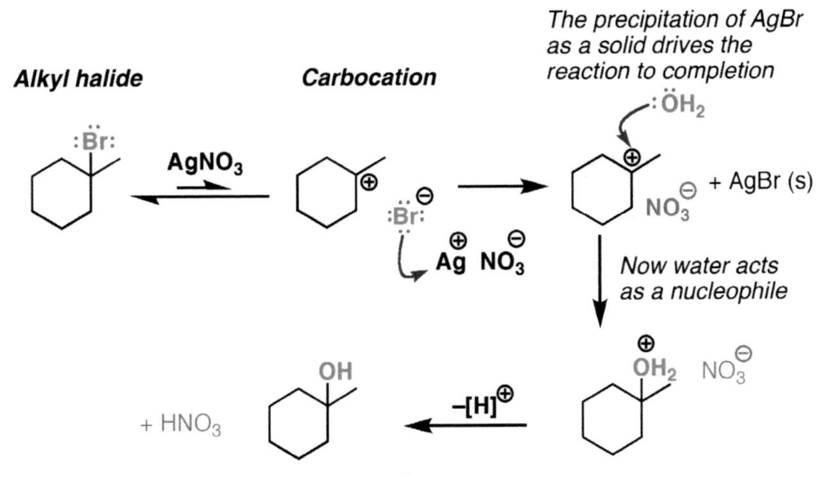

Either H_2O or $NO_3^{\ominus}$ could act as a base here.

Ag_2O
Silver Oxide

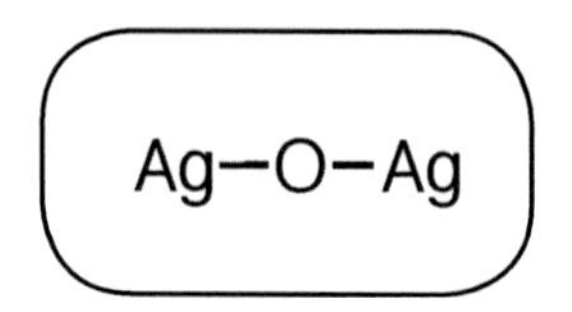

What it's used for: Silver oxide is used in the Tollens reaction to oxidize aldehydes to carboxylic acids. This is the basis of a test for the presence of aldehydes, since a mirror of Ag^{o} will be deposited on the flask. It is also used as the base in the Hoffman elimination.

Similar to: $AgNO_3$

Example 1: Tollens oxidation of aldehydes to carboxylic acids

Ag_2O, NH_3, H_2O + Ag (s)

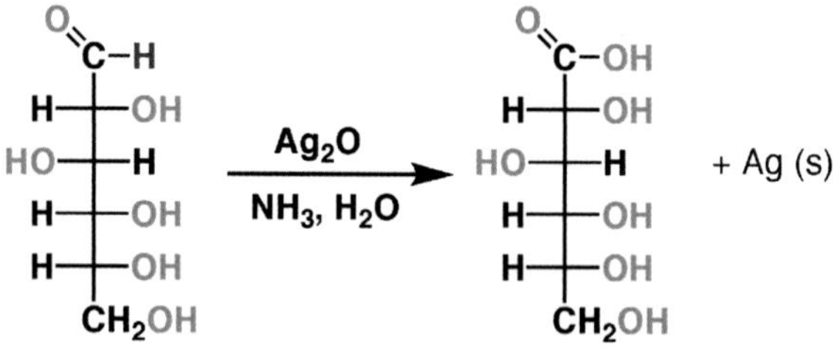

Aldehyde (open form)

This reaction is usually introduced in the context of sugar chemistry

Example 2: As the base in the Hoffmann elimination

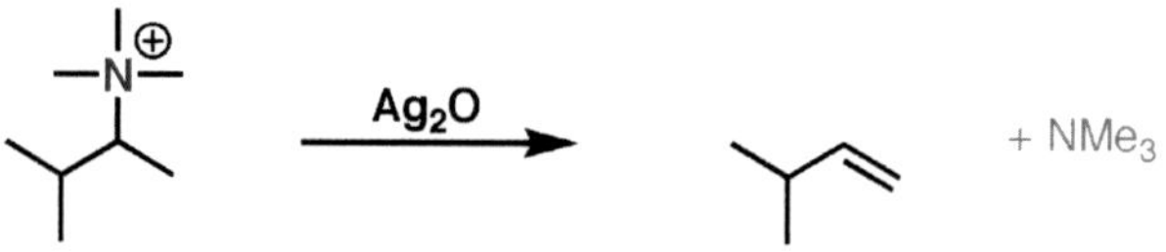

Ag_2O + NMe_3

note that the less substituted alkene is formed here

AIBN
[2,2’Azobis(2-methyl propionitrile)]

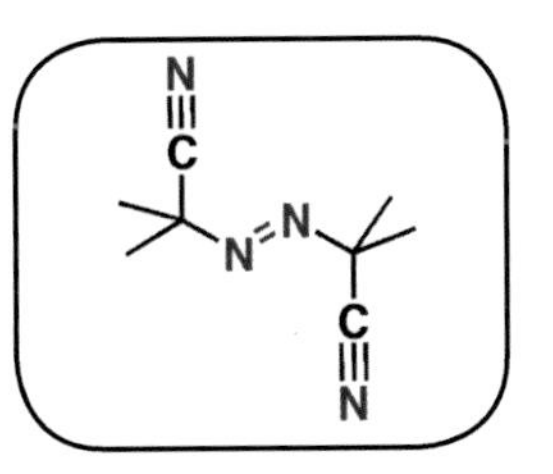

What it’s used for: Free radical initiator. Upon heating, AIBN decomposes to give nitrogen gas and two free radicals.

Similar to: RO–OR (“peroxides”) , benzoyl peroxide

Example 1: Free-radical halogenation of alkenes

AIBN, HBr
heat (Δ)
Br

How it works: ***Free-radical halogenation of alkenes***

AIBN heat

(2 equiv)

driving force for this reaction is the release of nitrogen gas

Initiation:

Br—H ·—C≡N ⟶ H—C≡N + Br ·

Propagation step 1:

· Br ⟶ Br

Propagation step 2:

Br HBr ⟶ H–Br Br ⟶ Br + Br ·

$AlBr_3$
Aluminum Bromide

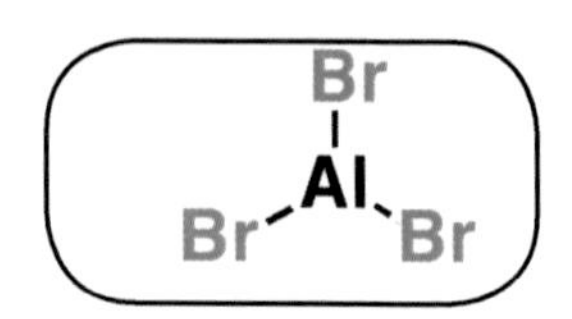

Also known as: Aluminum tribromide

What it's used for: Lewis acid, promoter for electrophilic aromatic substitution

Similar to: $FeCl_3$, $FeBr_3$, $AlCl_3$

Example 1: Electrophilic bromination - conversion of arenes to aryl halides

Benzene →(Br_2, $AlBr_3$)→ bromobenzene (Br) + $AlBr_3$ + HBr

Example 2: Friedel-Crafts acylation - conversion of arenes to aryl ketones

Benzene →(acetyl bromide: O, Br; $AlBr_3$)→ acetophenone (O) + $AlBr_3$ + HBr

Example 3: Friedel-Crafts alkylation - conversion of arenes to alkyl arenes

Benzene →(ethyl bromide: Br; $AlBr_3$)→ ethylbenzene + $AlBr_3$ + HBr

$AlCl_3$
Aluminum chloride

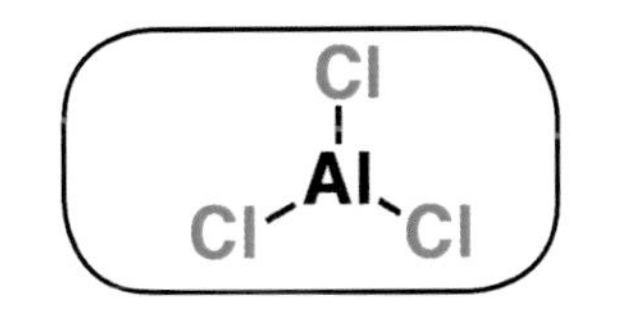

Also known as: Aluminum trichloride

What it's used for: Aluminum chloride is a strong Lewis acid. It can be used to catalyze the chlorination of aromatic compounds, as well as Friedel-Crafts reactions. It can also be used in the Meerwein-Ponndorf-Verley reduction.

Similar to: $AlBr_3$, $FeBr_3$, $FeCl_3$

Example 1: Electrophilic chlorination - conversion of arenes to aryl halides

Cl_2, $AlCl_3$ + $AlCl_3$ + HCl

Example 2: Friedel-Crafts acylation - conversion of arenes to aryl ketones

O, Cl, $AlCl_3$, O + $AlCl_3$ + HCl

Example 3: Friedel-Crafts alkylation - conversion of arenes to alkylarenes

Cl, $AlCl_3$ + $AlCl_3$ + HCl

Example 4: Meerwein-Ponndorf-Verley reduction - reduction of ketones and alcohols to aldehydes

O, $AlCl_3$, OH, OH + O, H

How it works: *Friedel-Crafts acylation*

O, R, Cl:, $AlCl_3$, :O:, R, ⊕ Cl:, ⊖ $AlCl_3$, O, R, ⊕, $AlCl_4^{\ominus}$

Acylium ion

O, ⊕, R, $AlCl_4^{\ominus}$, O, R, H, ⊕, ⊖ Cl—$AlCl_3$, O, R

+ $AlCl_3$ + HCl

$AlCl_3$ (continued)

How it works: ***Meerwein-Ponndorf-Verley Oxidation***

This reaction is typically run using an alcohol solvent such as ethanol or isopropanol. When $AlCl_3$ is added, the solvent replaces the chloro groups:

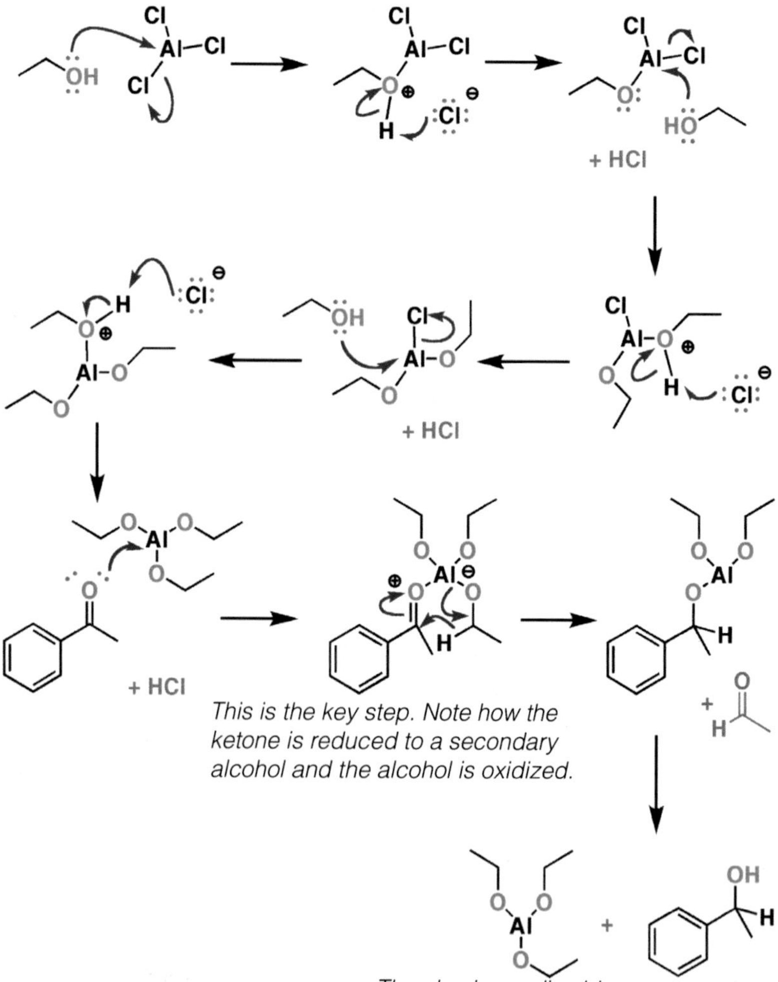

This is the key step. Note how the ketone is reduced to a secondary alcohol and the alcohol is oxidized.

The aluminum alkoxide can go on to catalyze further reactions.

BF_3
Boron Trifluoride

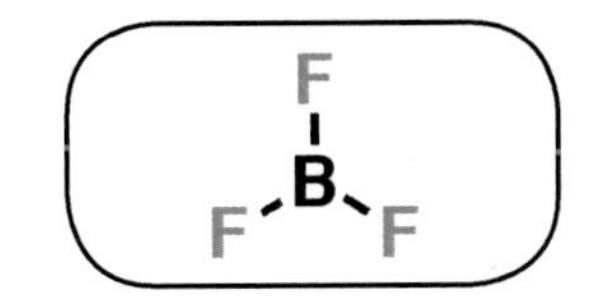

What it's used for: Boron trifluoride is a strong Lewis acid. It is commonly used for the formation of thioacetals from ketones (or aldehydes) with thiols. The product is a thioacetal.

Similar to: $FeCl_3$, $AlCl_3$ (also Lewis acids)

Example 1: Conversion of ketones to thioacetals

SH SH

BF_3

How it works: ***Formation of thioacetals***

BF_3 acts as a Lewis acid, coordinating to the carbonyl oxygen and activating the carbonyl carbon towards attack by sulfur.

Coordination

Addition

Proton transfer

Elimination

Addition

Deprotonation

Thioacetal

BH_3
Borane

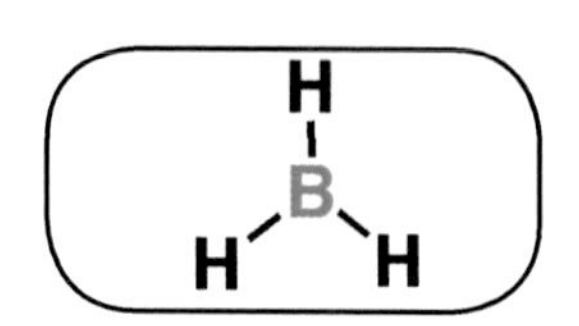

What it's used for: Borane is used for the hydroboration of alkenes and alkynes.

Similar to: B_2H_6 ("diborane"), BH_3•THF, BH_3•SMe_2, disiamylborane, 9-BBN (for our purposes, these can all be considered as "identical".

Example 1: Hydroboration reaction - conversion of alkenes to alcohols

1) BH_3
2) NaOH
H_2O_2

OH

1) BH_3
2) H_2O_2
NaOH

H
OH

note: stereochemistry is syn

Example 2: Hydroboration reaction - conversion of alkynes to aldehydes

H
1) BH_3
2) NaOH
H_2O_2

O
H

How it works: ***Hydroboration of alkenes***

Hydroboration is notable in that the **boron adds to the less substituted end of the alkene**. This is usually referred to as "anti-Marovnikoff" selectivity. The reason for the selectivity is that the **boron hydrogen bond is polarized so that the hydrogen has a partial negative charge and the boron has a partial positive charge (due to electronegativity)**. In the transition state, **the partially negative hydrogen "lines up" with the more substituted end of the double bond** (i.e. the end containing more bonds to carbon) since this will preferentially stabilize partial positive charge. The hydrogen and boron add **syn** to the double bond .

BH_3

δ⁻ δ⁺
H- -B H
H
δ⁺ δ⁻

H BH_2

syn addition

the transition state is concerted

If excess alkene is present, the two remaining B-H bonds can do subsequent hydroborations

BH_3 (continued)

The second step of the hydroboration is an oxidation that replaces the C–B bond with a C–O bond

$$\text{R–BH}_2 \xrightarrow{\text{NaOH, } H_2O_2} \text{R–OH} + H_2O + NaOH + HOBH_2$$

note retention of stereochemistry

The first step is deprotonation of hydrogen peroxide by sodium hydroxide; this makes the peroxide ion more nucleophilic (and more reactive)

$$\text{NaOH + HO–OH} \longrightarrow H_2O + \text{NaO–OH}$$

The deprotonated peroxide then attacks the boron, which then undergoes rearrangement to break the weak O–O bond. Then, hydroxide ion cleaves the B–O bond to give a deprotonated alcohol, which is then protonated by alcohol.

+ Na OH

How it works: ***Hydroboration of alkynes***

Hydroboration of alkynes forms a product called an ***enol***. Through a process called *tautomerism*, the enol product is converted into its more stable constitutional isomer, the **keto** form. In the case of a terminal alkyne (one which has a C-H bond) an aldehyde is formed.

1. BH_3
2. NaOH, H_2O_2

enol form (unstable)

keto form (more stable)

Br_2
Bromine

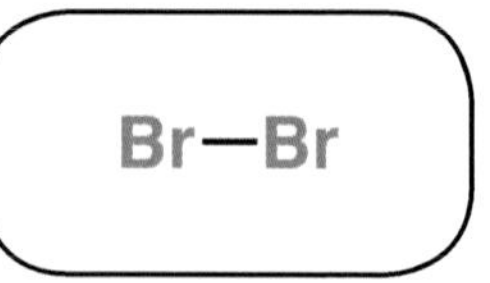

What it's used for: Bromine will react with alkenes, alkynes, aromatics, enols, and enolates, producing brominated compounds. In the presence of light, bromine will also replace hydrogen atoms in alkanes. Finally, bromine is also used to promote the Hoffmann rearrangement of amides to amines.

Similar to: NBS, Cl_2, I_2, NIS, NCS

Example 1: Bromination - conversion of alkenes to vicinal dibromides

Br_2 → Br, Br

Example 2: Bromination - conversion of alkynes to vicinal dibromides

Br_2 → Br, Br

Example 3: Conversion of alkenes to halohydrins

Br_2, H_2O → Br, OH + HBr

Example 4: Electrophilic bromination - conversion of arenes to aryl bromides.

Br_2, $FeBr_3$ *(or $AlBr_3$)* → Br + HBr

Example 5: Hoffmann rearrangement - conversion of amides to amines

O, NH_2 → 1) Br_2 2) H_2O → NH_2 + CO_2

Example 6: Conversion of ketones to α-bromoketones

O → $H^{\oplus}X^{\ominus}$, Br_2 → O, Br

"HX" is just a strong acid (e.g. HBr)

Example 7: Conversion of enolates to α-bromoketones

O → 1) LDA 2) Br_2 → O, Br + LiBr + HN

Br_2
(continued)

Example 8: Radical halogenation - conversion of alkanes to alkyl bromides

Br_2

light ($h\nu$)

+ HBr

Example 9: Haloform reaction - conversion of methyl ketones to carboxylic acids

NaOH

Br_2

OH

+ $HCBr_3$ + NaBr

How it works: ***Bromination of alkenes***

Br–Br

Br

:Br:

Br

Br

Bromonium ion

Treatment of an alkene with Br_2 leads to the formation of a bromonium ion, which undergoes backside attack. In the presence of a solvent that can act as a nucleophile, the halohydrin is obtained:

Br_2

H_2O

:Br:

Br

$:OH_2$

Br

OH

+ HBr

How it works: ***Bromination of alkenes***

Bromine is made more electrophilic by a Lewis acid such as $FeBr_3$; it can then undergo attack by an aromatic ring, resulting in electrophilic aromatic substitution of H for Br

Step 1: Activation

:Br–Br: $FeBr_3$ ⟶ :Br–Br–$FeBr_3$

Step 2: Electrophilic aromatic substitution

:Br–Br–$FeBr_3$

H

Br

Br–$FeBr_3$

Br

+ $FeBr_3$ + HBr

Br_2 (continued)

How it works: ***Hoffmann Rearrangement***

In this reaction, the lone pair on nitrogen attacks bromine, which leads to a rearrangement. Attack at the carbonyl carbon by water then leads to loss of CO_2, resulting in the formation of the free amine.

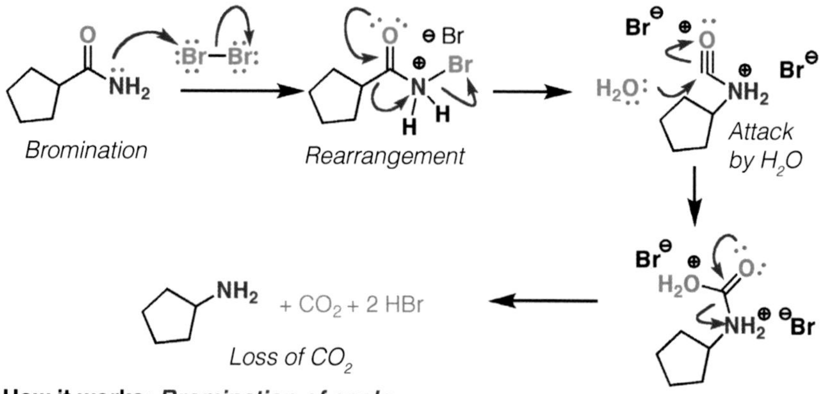

Bromination

Rearrangement

Attack by H_2O

+ CO_2 + 2 HBr

Loss of CO_2

How it works: ***Bromination of enols***

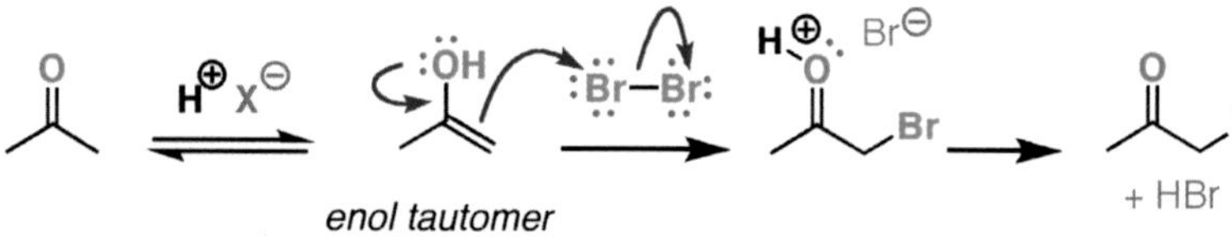

enol tautomer

+ HBr

How it works: ***Bromination of enolates***

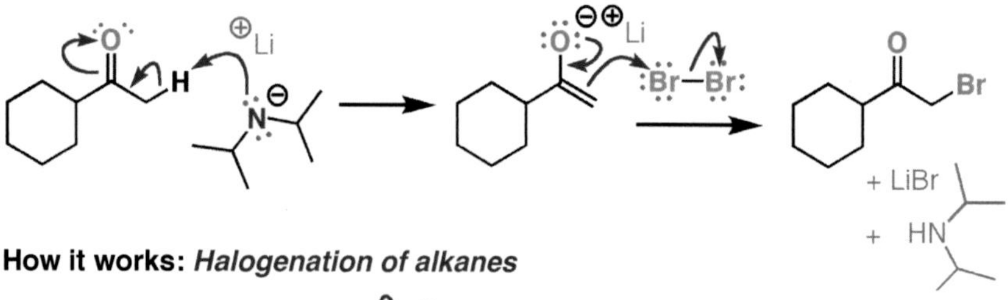

+ LiBr

+ HN

How it works: ***Halogenation of alkanes***

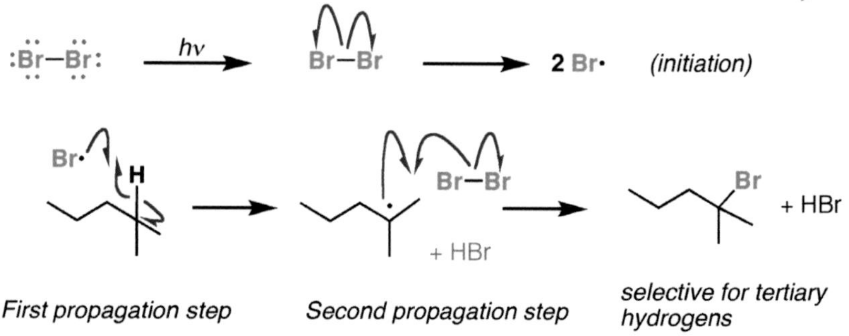

$h\nu$

2 Br· (initiation)

+ HBr

+ HBr

First propagation step

Second propagation step

selective for tertiary hydrogens

BsCl
p-bromobenzenesulfonyl chloride

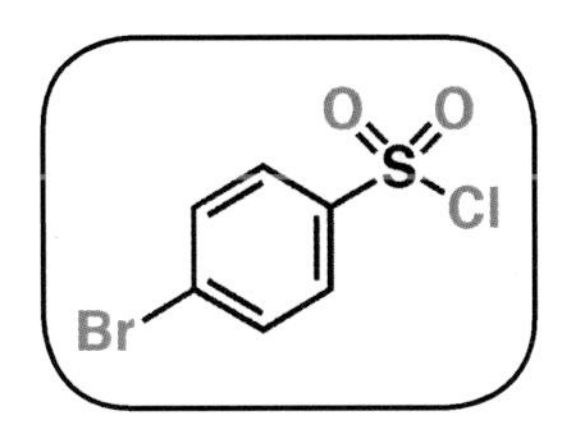

Also known as: Brosyl chloride

What it's used for: *p*-bromobenzene sulfonyl chloride (BsCl) is used to convert alcohols into good leaving groups. It is essentially interchangable with TsCl and MsCl for this purpose.

Similar to: TsCl, MsCl

Example 1: Conversion of alcohols into alkyl brosylates

OH → OBs

BsCl

pyridine (a weak base)

Cl_2
Chlorine

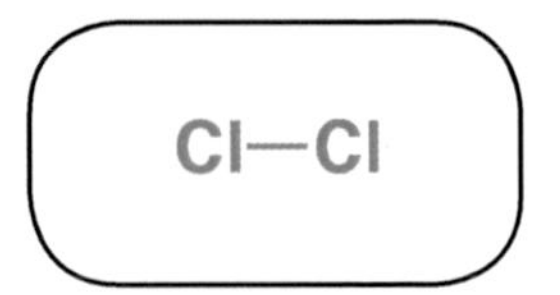

What it's used for: Chlorine is a very good electrophile. It will react with double and triple bonds, as well as aromatics, enols, and enolates to give chlorinated products. In addition it will substitute Cl for halogens when treated with light (free-radical conditions). Finally, it assists with the rearrangement of amides to amines (the Hoffmann rearrangement).

Similar to: NCS, Br_2, NBS, I_2, NIS

Example 1: Chlorination - conversion of alkenes to vicinal dichlorides

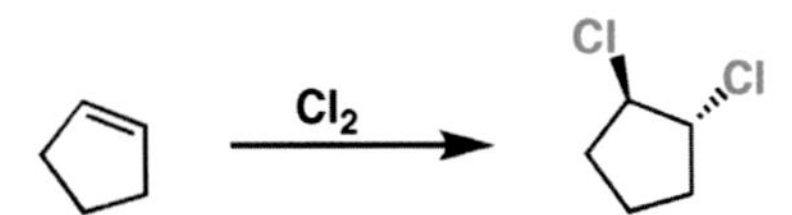

Example 2: Conversion of alkenes to chlorohydrins

Cl_2 / H_2O → Cl, OH + HCl

Example 3: Electrophilic chlorination - conversion of arenes to chloroarenes

Cl_2 / $AlCl_3$ (or $FeCl_3$) → Cl + HCl

Example 4: Hoffmann rearrangement - conversion of amides to amines

O, NH_2 — Cl_2 / $^{\ominus}OH$ → NH_2 + CO_2 + HCl

Example 5: Conversion of ketones to α-chloro ketones

O — $H^{\oplus}X^{\ominus}$ / Cl_2 → O, Cl

Example 6: Conversion ot enolates to α-chloro ketones

O — 1) LDA 2) Cl_2 → O, Cl + LiCl

HN

Example 7: Radical chlorination of alkanes to alkyl chlorides

Cl_2 / light (hν) → Cl + HCl

Cl_2 (continued)

Example 8: The haloform reaction

NaOH, Cl_2; OH; + $HCCl_3$ + NaCl

How it works: ***Chlorination of alkenes***

Cl–Cl; Cl; :Cl:; Cl; Cl

Chloronium ion

Note how the anti product is formed exclusively, through backside attack on the chloronium ion

How it works: ***Chlorohydrin formation***

Cl_2, H_2O; :Cl:; Cl; :OH_2; Cl; OH; + HCl

In a nucleophilic solvent such as H_2O, water will attack the chloronium ion, forming a chlorohydrin

How it works: ***Electrophilic chlorination***

Step 1: Activation

:Cl–Cl: $FeCl_3$ → :Cl–Cl—$FeCl_3$

In the first step of this reaction, a Lewis acid such as $FeCl_3$ activates Cl_2 towards attack by the aromatic ring.

Step 2: Electrophilic aromatic substitution

:Cl–Cl—$FeCl_3$ → H, Cl, Cl–$FeCl_3$ → Cl

In the second step, electrophilic aromatic substitution results in replacement of C–H by C-Cl

+ $FeCl_3$ + HCl

How it works: ***Chlorination of ketones under acidic conditions***

O; $H^{\oplus}X^{\ominus}$; :OH; :Cl–Cl:; H, O, $Cl^{\ominus}$; Cl; O; Cl; + HCl

enol tautomer

Cl_2 (continued)

How it works: ***Chlorination of enolates***

Deprotonation of the ketone by strong base results in an enolate, which then attacks Cl_2

+ LiCl + HN

How it works: ***Chlorination of alkanes***

Initiation step

:Cl–Cl: —hν→ :Cl–Cl: → 2 Cl•

Propagation step #1

Cl• → + HCl

Propagation step #2

+ Cl•

How it works: ***Hoffmann Rearrangement***

+ CO_2 + 2 HCl

CN
Cyanide ion

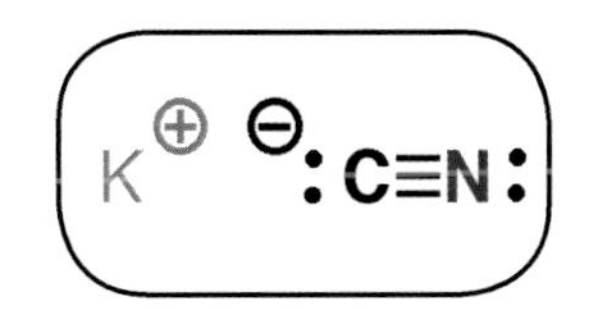

What it's used for: Cyanide ion is a good nucleophile. It can be used for substitution reactions (S_N2), for forming cyanohydrins from aldehydes or ketones, and in the benzoin condensation.

Same as: KCN, NaCN, LiCN

Example 1: As a nucleophile in substitution reactions

Br —(KCN, DMSO)→ CN + KBr

Example 2: Formation of cyanohydrins from aldehydes/ketones

O, H —(KCN, H⊕)→ HO, C≡N, H

O —(KCN, H⊕)→ HO, C≡N

Example 3: In the benzoin condensation

(2 equiv) O, H —(KCN)→ OH, O *Benzoin*

How it works: ***Nucleophilic substitution***

N≡C:⊖ ⊕K, Br → CN + KBr

Cyanide ion is a good nucleophile but a weak base (pKa of 9)

How it works: ***Benzoin condensation***

O, H, :C≡N ⊖, K⊕ → K⊕ :O:⊖, H, C≡N ⇌ :OH, ⊖K⊕, C≡N; O, H

Here, the proton is transferred betweeen carbon and oxygen

Carbonyl addition

H, :O:, :O:⊖ K⊕, C≡N, H

Expulsion of cyanide

+ KCN ← O, :OH, H

CrO_3
Chromium trioxide

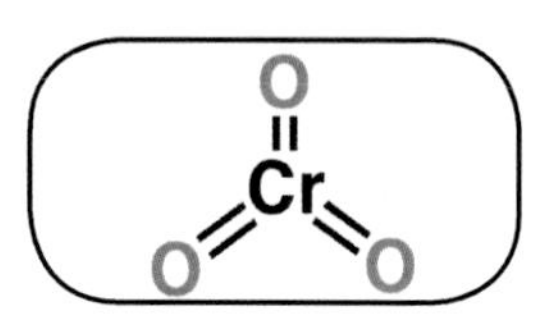

What it's used for: CrO_3 is an oxidant. When pyridine is present, it is a mild oxidant that will oxidize primary alcohols to aldehydes. However, if water and acid are present, the aldehyde will be oxidized further the the carboxylic acid.

Similar to: PCC (when pyridine is added)
When aqueous acid is present, it is the same or similar to Na_2CrO_4 / $K_2Cr_2O_7$ / $Na_2Cr_2O_7$ / H_2CrO_4 (and $KMnO_4$). Watch out! this reagent is the source of much confusion!

Example 1: Oxidation of primary alcohols to aldehydes (with pyridine)

OH → CrO_3 / (pyridine) → O, H

Example 2: Oxidation of secondary alcohols to ketones (with pyridine)

OH → CrO_3 / (pyridine) → O

Example 3: Oxidation of primary alcohols to carboxylic acids

OH → CrO_3 / $H_2O/H_3O^{\oplus}$ → O, OH

How it works: ***Oxidation of primary alcohols to aldehydes***

proton transfer

pyridine (a base)

+ $HCrO_3^{\ominus}$

CrO_3
(continued)

How it works: ***Oxidation of primary alcohols to carboxylic acids***

When water is present the aldehyde will form the hydrate, which will be further oxidized to the carboxylic acid.

O OH Cr H H H O O ⊕ X⊖

$H_2\ddot{O}$:

Water is a strong enough base to deprotonate here

O H

$H_2\ddot{O}$:

H_2O

OH H O H O Cr=O O

Hydrate

OH ⊕ H H O O⊖ Cr O O

$H_2\ddot{O}$:

Second deprotonation results in formation of the carbonyl

OH ⊕ O H O⊖ Cr O O

O OH

\+ $HCrO_3^-$

After proton transfer

CuBr
Copper (I) Bromide

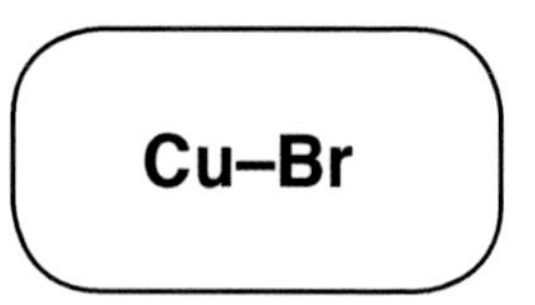

Also known as: Cuprous bromide

What it's used for: Reacts with aromatic diazonium salts to give aromatic bromides. Also used to make organocuprates (Gilman reagents).

Similar to: Copper(I) cyanide (CuCN), Copper(I) chloride, Copper(I) iodide

Example 1: Formation of aryl bromides from aryl diazonium salts

$PhN_2^+ X^- \xrightarrow{CuBr} PhBr + CuX + N_2$

Example 2: Formation of organocuprate reagents (Gilman reagents)

(2 equiv) $BuLi \xrightarrow{CuBr} Bu_2Cu^- Li^+ + LiBr$

How it works: ***Formation of aryl bromides***

Not perfectly understood!
It is known that this reaction occurs through a free radical process. Here is a suggested mechanism:

Donation of an electron by Cu(I) to give Cu(II)

Driving force for this reaction is loss of nitrogen gas!

$-N_2$

+ CuBr

The radical then abstracts Br from $CuBr_2$, giving CuBr

CuCl
Copper (I) Chloride

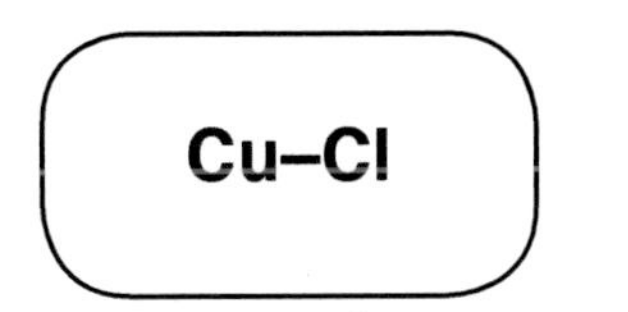

Also known as: Cuprous chloride

What it's used for: Reacts with aromatic diazonium salts to give aryl chlorides; also used to form organocuprates (Gilman reagents) from organolithium salts.

Similar to: Copper(I) cyanide (CuCN), Copper bromide, Copper Iodide

Example 1: Formation of aryl chlorides from diazonium salts

$PhN_2^+ X^- \xrightarrow{CuCl} PhCl + CuX + N_2$

Example 2: Formation of organocuprates (Gilman reagents)

$BuLi$ (2 equiv) $\xrightarrow{CuCl} Bu_2Cu^- Li^+ + LiCl$

How it works: ***Formation of aryl chlorides from aryl diazonium salts***

Not perfectly understood, although proceeds through a free radical process.
Suggested mechanism:

Donation of an electron by Cu(I) to give Cu(II)

Driving force for this reaction is loss of nitrogen gas! $-N_2$

The radical then abstracts Cl from $CuCl_2$, giving CuCl

CuI
Copper (I) Iodide

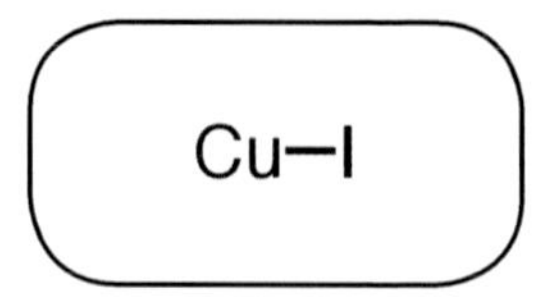

Also known as: Cuprous iodide

What it's used for: Reacts with alkyllithium reagents to form dialkyl cuprates

Similar to: CuBr, CuCN, CuCl

Example 1: Formation of dialkyl cuprates (Gilman reagents)

Li (2 equiv) —CuI→ Cu⊖ Li⊕ + LiI

How it works: ***Formation of organocuprates***

Li + CuI → Cu + LiI

Cu + Li → Cu⊖ Li⊕

Cuprates can be used to do conjugate additions [1,4 addition]:

O ... Cu⊖ Li⊕ → :O:⊖ Li⊕ + Cu —$H_3O^{\oplus}X^{\ominus}$→ O + LiX

They will also add to acyl halides to give ketones:

O Cl, Cu⊖ Li⊕ → ⊖ :O: Li⊕ Cl + Cu → O + LiCl

DCC
N,N'-dicyclohexane carbodiimide

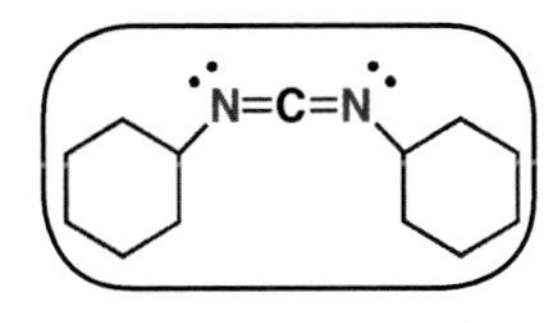

What it's used for: DCC is primarily used for the synthesis of amides from amines and carboxylic acids. It is, essentially, a dehydration reagent (removes water)

Example 1: Formation of amides from carboxylic acids and amines

H_2N / DCC

How it works: ***Formation of amides from carboxylic acids and amines***

The first step is attack of the carbon on the imide by the oxygen on the carboxylic acid.

Proton transfer

Now the amine attacks!

Amide

This byproduct is called a "urea" (formed after proton transfer)

DMS
Dimethyl sulfide

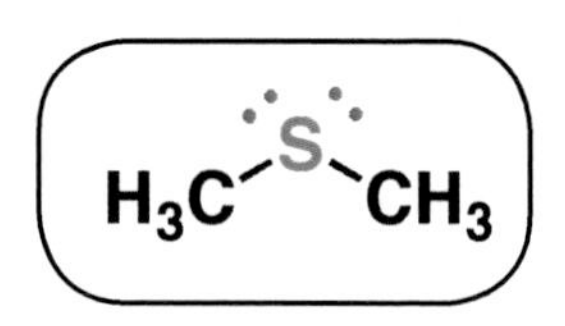

Also known as: Me_2S, methyl sulfide

What it's used for: Used in the "reductive workup" of ozonolysis, to reduce the ozonide that is formed. DMS is oxidized to dimethyl sulfoxide (DMSO) in the process.

Similar to: Zn (in the reductive workup for ozonolysis)

Example 1: Reductive workup for ozonolysis

1) O_3
2) S

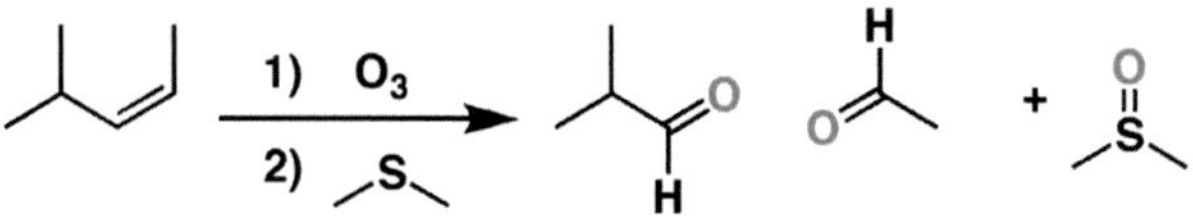

How it works: ***Reductive workup for ozonolysis***

The first step is formation of an ozonide by treating an alkene with O_3

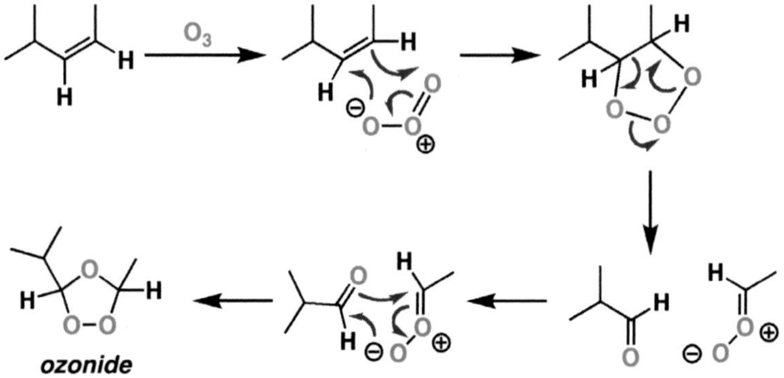

ozonide

In the second step, the ozonide is treated with DMS, which results in reduction of the ozonide and formation of dimethyl sulfoxide (DMSO)

SMe_2

DMSO

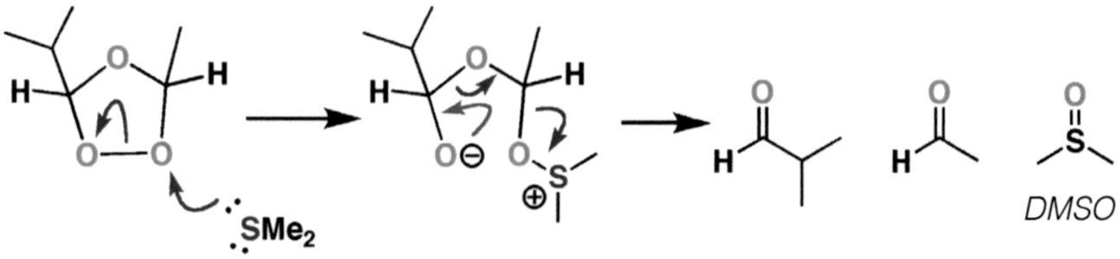

CH_2N_2
Diazomethane

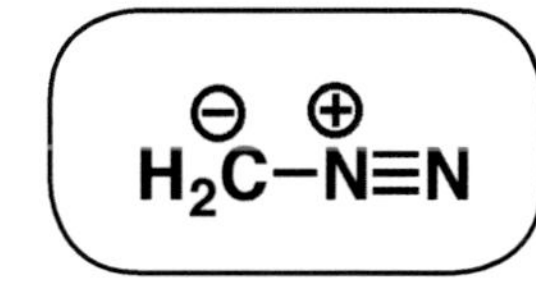

What it's used for: Diazomethane is used for three main purposes: 1) to convert carboxylic acids into methyl esters, and 2) in the Wolff rearrangement, as a means to extend carboxylic acids by one carbon, and 3) for cyclopropanation of alkenes.

Example 1: Conversion of carboxylic acids to methyl esters

CH_2N_2 + N_2

Example 2: Cyclopropanation of alkenes

CH_2N_2 + N_2

Example 3: In the Wolff Rearrangement

1) CH_2N_2
2) heat (Δ)
3) H_2O

How it works: ***Formation of methyl esters***

$H_2C-N\equiv N$ → $H_3C-N\equiv N$ → CH_3 + N_2

How it works: ***Wolff Rearrangement***

Step 1 is addition of diazomethane to the acid choride and displacement of Cl.

Addition *Eliimination* + HCl

Step 2 is heat, which initiates the rearrangement, forming a ketene.

heat *Heating leads to loss of N_2 gas* ***ketene***

In step 3, addition of water forms the carboxylic acid.

Tautomerism *(after proton transfer)*

D
Deuterium

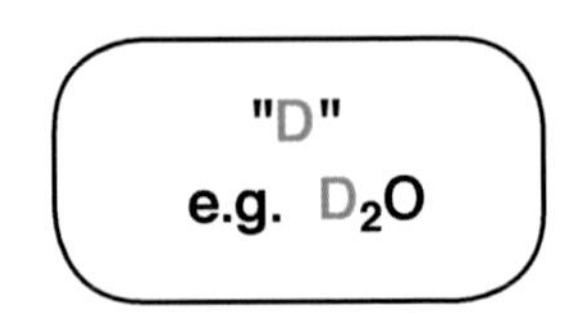

Also known as: "heavy hydrogen"

What it's used for: Deuterium is the heavy isotope of hydrogen, having an atomic weight of two. Deuterium has essentially the same reactivity as hydrogen, but due to the different magnetic properties of the nucleus, it can be differentiated from hydrogen in ^1H NMR. Deuterium analogs of hydrogen-containing reagents can therefore be useful in introducing deuterium as a "label" for examining stereochemistry and mechanisms.

Example 1: Deuterium reagents as acids

PhMgBr —(D_2O)→ PhD

$H_3C—C≡C—H$ —(1) $NaNH_2$ 2) D_2O)→ $H_3C—C≡C—D$

Example 2: Hydroboration of alkenes

1-methylcyclohexene —(1) BD_3 2) H_2O_2, NaOH)→ product with D and OH

note: stereochemistry is syn

Example 3: Reduction of ketones

cyclohexanone —($NaBD_4$)→ cyclohexanol with HO and D on same carbon

How it works: ***Deuterium as a reagent***

For examples of the mechanisms, see the section for the corresponding hydrogen reagents.

DIBAL

Di-isobutyl aluminum hydride

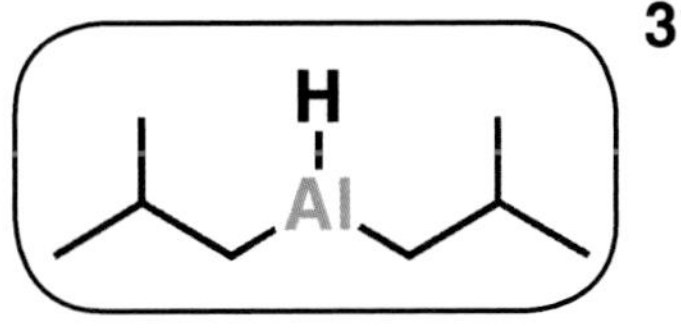

What it's used for: Strong, bulky reducing agent. It is most useful for the reduction of esters to aldehydes: unlike $LiAlH_4$, it will not reduce the aldehyde further unless an extra equivalent is added. It will also reduce other carbonyl compounds such as amides, aldehydes, ketones, and nitriles.

Similar to: $LiAlH_4$ (LAH), $LiAlH(Ot\text{-}Bu)_3$

Example 1: Reduction of esters to aldehydes

MeO(C=O)CH(CH₃)₂ → 1) DIBAL-H, –70°C; 2) H_2O → H(C=O)CH(CH₃)₂

Low temperature is important to prevent further reduction

Example 2: Reduction of ketones to secondary alcohols

Ketone → 1) DIBAL-H; 2) H_2O → secondary alcohol (H, OH)

Example 3: Reduction of aldehydes to primary alcohols

Aldehyde → 1) DIBAL-H; 2) H_2O → primary alcohol (H, H, OH)

Example 4: Reduction of nitriles to aldehydes

Nitrile (C≡N) → 1) DIBAL-H; 2) H_2O/HCl → aldehyde (C(=O)H)

The reaction initially forms an imine, which is then hydrolyzed by acid

Example 5: Reduction of acyl halides to aldehydes

Cl(C=O)CH(CH₃)₂ → 1. DIBAL-H, Hexane –70°C; 2. H_2O → H(C=O)CH(CH₃)₂

Low temperature is important to prevent further reduction

DIBAL

(continued)

How it works: ***Reduction of esters to aldehydes***

With its bulky isobutyl groups, DIBAL is more sterically hindered than $LiAlH_4$. If the temperature is kept low, DIBAL can reduce an ester to an aldehyde without subsequent reduction to the alcohol.

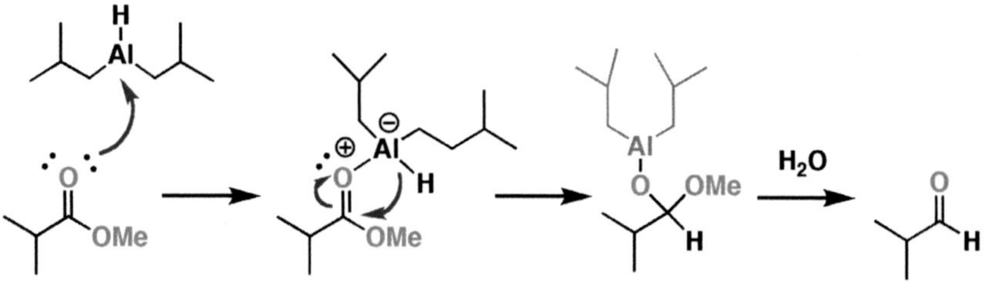

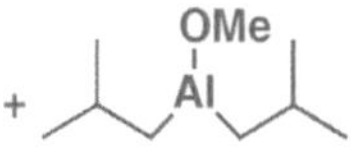

The first step is coordination of the oxygen lone pair to the aluminum

Next, hydride is delivered to the carbonyl carbon

At low temperatures the product is stable until acid or water is added to quench.

How it works: ***Reduction of esters to aldehydes***

Coordination of the nitrogen lone pair to the aluminum

Delivery of hydride to the nitrile carbon

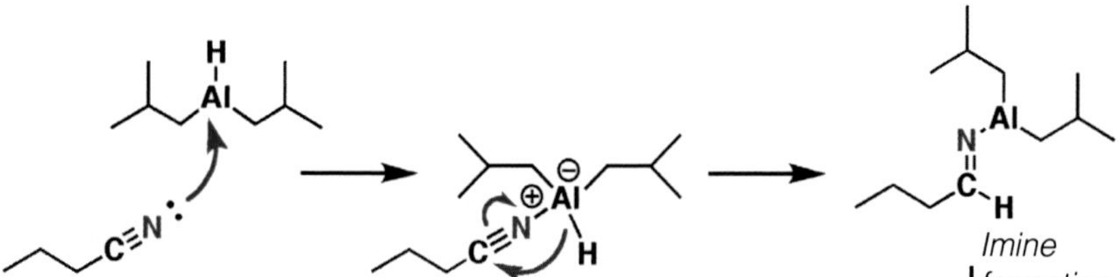

Imine formation

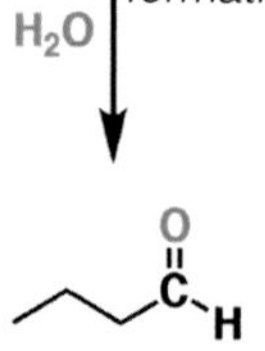

+ NH_3
Hydrolysis gives an aldehyde

DMP
Dess-Martin Periodinane

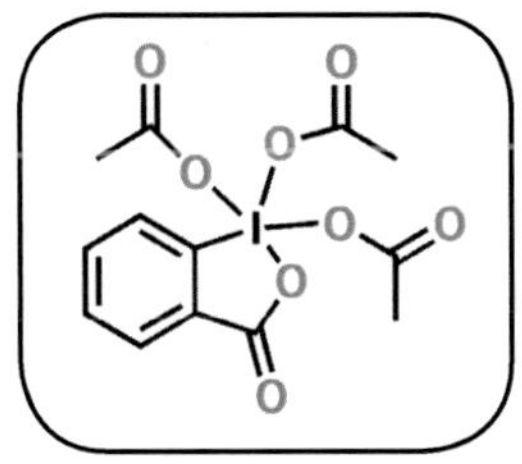

What it's used for: Dess-Martin periodinane is an oxidizing agent. It will oxidize primary alcohols to aldehydes without going to the carboxylic acid (similar to PCC). It will also oxidize secondary alcohols to ketones.

Similar to: PCC, CrO_3 with pyridine

Example 1: Oxidation - conversion of primary alcohols to aldehydes

OH H H —Dess-Martin→ O H

Example 2: Oxidation - conversion of secondary alcohols to ketones

HO H —Dess-Martin→ O

How it works: ***Oxidation of alcohols***

The mechanism for oxidation of alcohols by Dess-Martin periodinane is almost never covered in introductory textbooks. However it is included here in the interests of completeness. Mechanism is the same for primary and secondary alcohols.

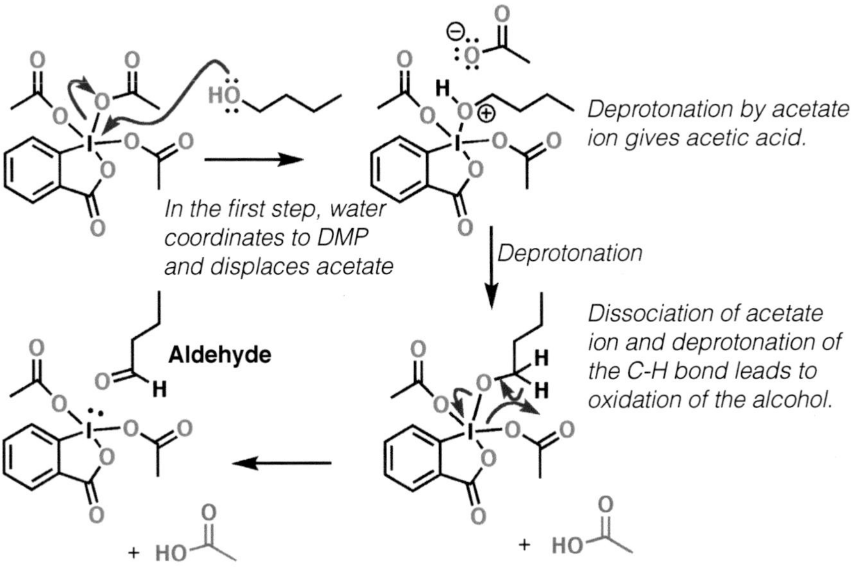

In the first step, water coordinates to DMP and displaces acetate

Deprotonation by acetate ion gives acetic acid.

Deprotonation

Dissociation of acetate ion and deprotonation of the C-H bond leads to oxidation of the alcohol.

Aldehyde

+ HO (acetic acid) + HO (acetic acid)

Fe
Iron

Fe

What it's used for: Iron metal (Fe) will reduce nitro groups to amines in the presence of a strong acid such as HCl.

Similar to: Tin (Sn), zinc (Zn)

Example 1: Reduction: conversion of nitro groups to primary amines

NO_2 — Fe / HCl → NH_2 + $FeCl_2$ + H_2O

How it works: ***Reduction of nitro groups***

The mechanism for this reaction is complex and proceeds in multiple steps. It likely proceeds similarly to that drawn in the section for tin.

$FeBr_3$
Iron (III) Bromide

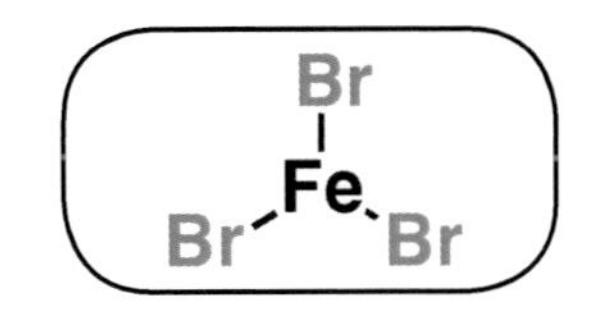

Also known as: Ferric bromide, iron tribromide

What it's used for: Lewis acid, promoter for electrophilic aromatic substitution
Similar to: $AlBr_3$, $AlCl_3$, $FeCl_3$

Example 1: Electrophilic bromination - conversion of arenes to aryl bromides

Benzene —(Br_2, $FeBr_3$)→ bromobenzene + HBr + $FeBr_3$

Example 2: Friedel-Crafts acylation - conversion of arenes to aryl ketones

Benzene —(acetyl bromide, $FeBr_3$)→ acetophenone + HBr + $FeBr_3$

Example 3: Friedel-Crafts alkylation - conversion of arenes to alkyl arenes

Benzene —(ethyl bromide, $FeBr_3$)→ ethylbenzene + HBr + $FeBr_3$

How it works: ***Electrophilic bromination***

$FeBr_3$ is a Lewis acid that can coordinate to halogens. In doing so it increases their electrophilicity, making them much more reactive.

:Br–Br: + $FeBr_3$ → $Br^{\oplus}$ $FeBr_4^{\ominus}$

This is a more electrophilic source of bromine than Br_2

Benzene + $Br^{\oplus}$ $FeBr_4^{\ominus}$ → arenium ion (H, Br) + $Br–FeBr_3^{\ominus}$ → bromobenzene + $FeBr_3$ + HBr

Trivia: $FeBr_3$ can also be used for chlorination, but $FeCl_3$ is more often used. The reason is that small amounts of halide scrambling can occur when $FeBr_3$ is used with Cl_2

Benzene —(Cl_2, $FeBr_3$)→ chlorobenzene (***major***) + bromobenzene (***minor***)

$FeBr_3$ (continued)

How it works: ***Friedel-Crafts Acylation***

Coordination of the Lewis acid $FeBr_3$ to the Br of the acid halide makes Br a better leaving group, facilitating formation of the carbocation ("acylium ion" in this case).

$FeBr_3$

$\oplus$ $\ominus$ Br–$FeBr_3$

$\oplus$ $\ominus FeBr_4$

Acylium ion

Next, attack of the aromatic ring upon the carbocation followed by deprotonation gives the aryl ketone.

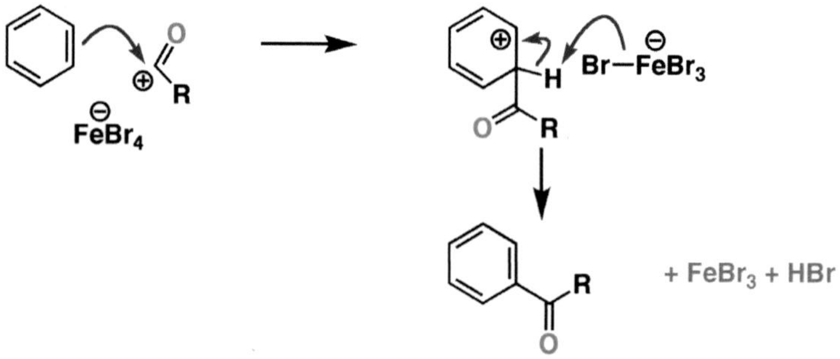

$\ominus FeBr_4$

$\ominus$ Br—$FeBr_3$

+ $FeBr_3$ + HBr

A similar process operates for the Friedel-Crafts alkylation (not pictured)

$FeCl_3$
Iron (III) chloride

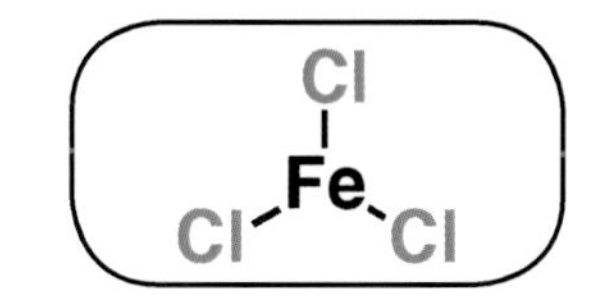

Also known as: Ferric chloride, iron trichloride

What it's used for: Iron (III) chloride (ferric chloride) is a Lewis acid. It is useful in promoting the chlorination of aromatic compounds with Cl2 as well as in the Friedel-Crafts alkylation and acylation reactions.

Similar to: $AlCl_3$, $AlBr_3$, $FeBr_3$

Example 1: Electrophilic chlorination - conversion of arenes to aryl chlorides

Cl_2, $FeCl_3$ → Cl + HCl + $FeCl_3$

Example 2: Friedel-Crafts acylation - conversion of arenes to aryl ketones

O, Cl, $FeCl_3$ → O + HCl + $FeCl_3$

Example 3: Friedel-Crafts alkylation: conversion of arenes to alkylarenes

Cl, $FeCl_3$ → + HCl + $FeCl_3$

How it works:

See sections on $AlCl_3$ and $FeBr_3$ - $FeCl_3$ works in exactly the same way.

Grignard Reagents

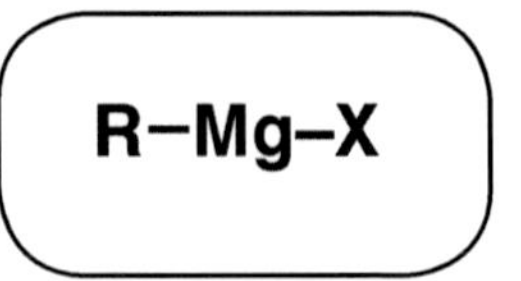

Also known as: Organomagnesium reagents

What it's used for: Extremely good nucleophile, reacts with electrophiles such as carbonyl compounds (aldehydes, ketones, esters, carbon dioxide, etc.) and epoxides. In addition Grignard reagents are very strong bases and will react with acidic hydrogens.

Similar to: Organolithium reagents (R–Li)

Example 1: Conversion of alkyl or alkenyl halides to Grignard reagents

Cl → Mg → MgCl

Grignards can be formed from alkyl or alkenyl chlorides, bromides, or iodides (never fluorides)

Br → Mg → MgBr

Example 2: Conversion of aldehydes to secondary alcohols

O, H → 1) MgCl 2) $H_3O^{\oplus}X^{\ominus}$ → OH + MgClX

Acid is added in the second step to protonate the negatively charged oxygen.

Example 3: Conversion of ketones to tertiary alcohols

O → 1) MgCl 2) $H_3O^{\oplus}X^{\ominus}$ → OH + MgClX

Example 4: Conversion of esters to tertiary alcohols

O, O → 1) MgCl 2) $H_3O^{\oplus}X^{\ominus}$ → OH + HO + 2 MgClX

Grignard reagents add twice to esters, acid halides, and anhydrides

Example 5: Conversion of acyl halides to tertiary alcohols

O, Cl → 1) MgBr 2) $H_3O^{\oplus}X^{\ominus}$ → OH + 2 MgBrX

Grignard reagents
(continued)

Example 6: Reaction with epoxides

1) BrMg
2) $H_3O^{\oplus}X^{\ominus}$
HO + MgBrX

Grignard reagents add to the less substituted end of epoxides

Example 7: Reaction with carbon dioxide

MgBr 1) CO_2 2) $H_3O^{\oplus}X^{\ominus}$ CO_2H + MgBrX

The purpsoe of acid in the second step is to protonate the negatively charged oxygen.

Example 8: Reaction with acidic hydrogens

O OH MgBr O $O^{\ominus}$ $^{\oplus}$MgBr + H

This can be used to introduce deuterium:

MgBr D_2O D + $DO^{\ominus}$ $^{\oplus}$MgBr

Deuterium is the heavy isotope of hydrogen

How it works: ***Addition to aldehydes and ketones***

Grignard reagents are extremely strong nucleophiles. The electrons in the C–Mg bond are heavily polarized towards carbon

δ^- δ^+ MgBr essentially behaves like $\ominus$ $\oplus$ MgBr

Therefore, Grignard reagents will react well with electrophiles such as aldehydes and ketones.

O H MgBr [Addition] $H^{\oplus}X^{\ominus}$ $O^{\ominus}$ $^{\oplus}$MgBr H acid OH + MgBrX

Acid is added after completion of the addition step

Grignard Reagents
(continued)

How it works: ***Addition to epoxides***

+ MgClX

How it works: ***Addition to epoxides***

These proceed through a two step mechanism: addition followed by elimination. Acid is added at the end to obtain the alcohol.

Addition of Grignard reagent to the ester

Elimination of the OR group then forms the ketone

A second equivalent of Grignard reagent then adds to the ketone

Finally, acid [HX here] is added to obtain the neutral alcohol

+ MgClX

The same mechanism operates for acid halides and anhydrides.

H_2
Hydrogen

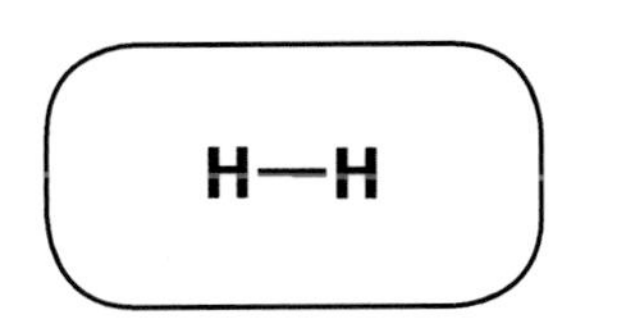

What it's used for: Hydrogen gas is used for the reduction of alkenes, alkynes, and many other species with multiple bonds, in concert with catalysts such as Pd/C and Pt.

Example 1: Hydrogenation - conversion of alkenes to alkanes

H_2, Pd/C — H, H — *syn addition*

Example 2: Hydrogenation - conversion of alkynes to alkanes

H_2, Pd/C → CH_2–CH_2

Example 3: Lindlar reduction - conversion of alkynes to alkenes

Lindlar's catalyst, H_2 → H, H

Example 4: Reduction - conversion of nitro groups to primary amines

NO_2 → (H_2, Pd/C) → NH_2

Example 5: Hydrogenation - conversion of nitriles to primary amines

C≡N → (H_2, Pd/C) → CH_2NH_2

Example 6: Hydrogenation - conversion of imines to amines

N–R → (H_2, Pd/C) → H HN–R

Example 7: Hydrogenation - conversion of arenes to cycloalkanes

Pd/C, H_2, heat (Δ), high pressure

HBr
Hydrobromic acid

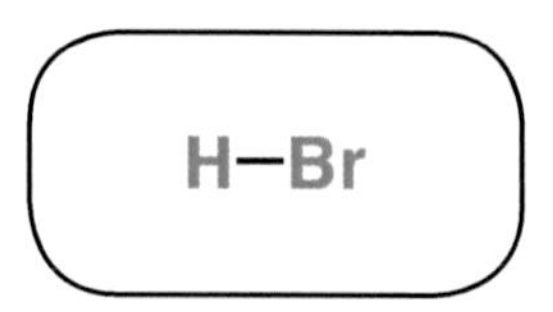
H–Br

What it's used for: Hydrobromic acid is a strong acid. It can add to compounds with multiple bonds such as alkenes and alkynes. It can also react with primary, secondary, and tertiary alcohols to form alkyl bromides.

Similar to: HCl, HI

Example 1: Hydrohalogenation - conversion of alkenes to alkyl bromides

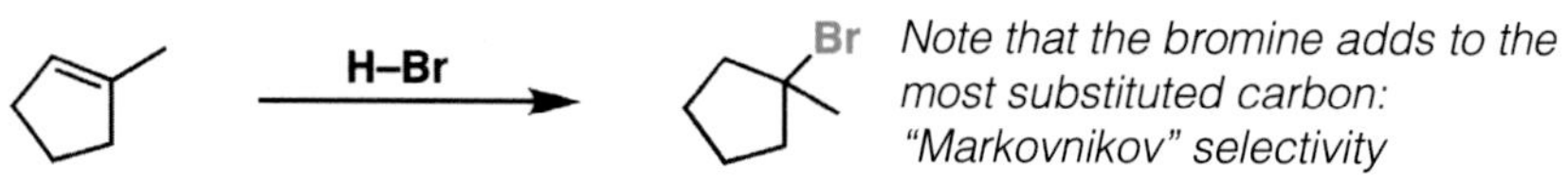
H–Br
Br
Note that the bromine adds to the most substituted carbon: "Markovnikov" selectivity

Example 2: Hydrohalogenation - conversion of alkynes to alkenyl bromides

H
H–Br
(1 equiv)
Br

Example 3: Hydrohalogenation - conversion of alkynes to geminal dibromides

H
H–Br
(2 equiv)
Br Br
"geminal" dibromide

Example 4: Free-radical addition - conversion of alkenes to alkyl bromides

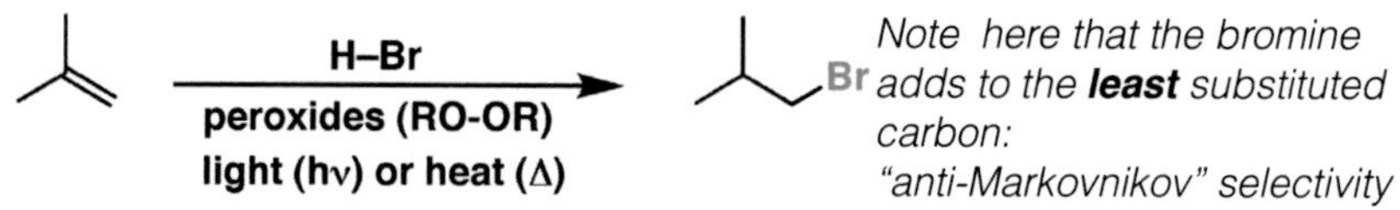
H–Br
peroxides (RO-OR)
light (hν) or heat (Δ)
Br
Note here that the bromine adds to the least substituted carbon:
"anti-Markovnikov" selectivity

Example 5: Conversion of alcohols to alkyl bromides (S_N2)

OH
H–Br
Br + H_2O
Primary alcohol, hence S_N2 here.

Note: occurs through S_N2

Example 6: Conversion of alcohols to alkyl bromides (S_N1)

OH
H–Br
Br + H_2O
Tertiary alcohol, hence S_N1

Note: occurs through S_N1

How it works: ***Addition to alkenes***

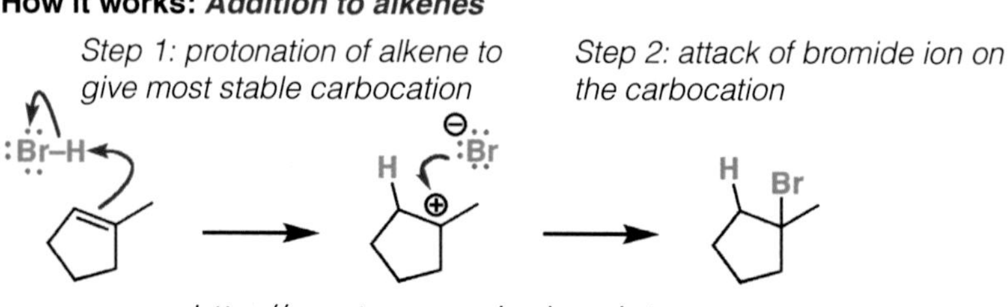
Step 1: protonation of alkene to give most stable carbocation
Step 2: attack of bromide ion on the carbocation
:Br–H
H
⊖
:Br
⊕
H
Br

HBr
(continued)

How it works: ***Addition to alkynes***
Addition of 1 equivalent of HBr will lead to a vinyl bromide; addition of a second equivalent leads to the geminal dibromide

Attack of bromide upon carbocation

H–Br

Formation of most stable carbocation

H–Br

Attack of bromide upon carbocation

H–Br

Formation of most stable carbocation

How it works: ***Formation of alkyl bromides from alcohols***
Protonation of OH by HBr makes a good leaving group (H_2O). When a stable carbocation cannot be formed, the reaction proceeds via an S_N2 pathway:

H–Br *protonation* *backside attack* + H_2O

Tertiary alcohols tend to proceed through an SN1 pathway:

H–Br *protonation* *attack of bromide ion* + H_2O

How it works: ***Free radical addition of HBr to alkenes***
Peroxides (general formula RO-OR) have a weak O–O bond and will fragment homolytically upon treatment with heat or light to give peroxy radicals:

RO—OR *heat or light* → 2 RO·

Peroxy radicals are very reactive; they will readily remove hydrogen from various groups (e.g. HBr) giving rise to free radical chain processes:

RO· + H–Br → ROH + ·Br

only a catalytic amount of peroxides are required to initiate the reaction

Propagation step 1:

·Br

Here, bromine radical adds to the alkene. Note that addition occurs at the less substituted carbon; this gives rise to the most stable free radical (secondary in this case)

Propagation step 2:

H–Br → + ·Br

HCl
Hydrochloric acid

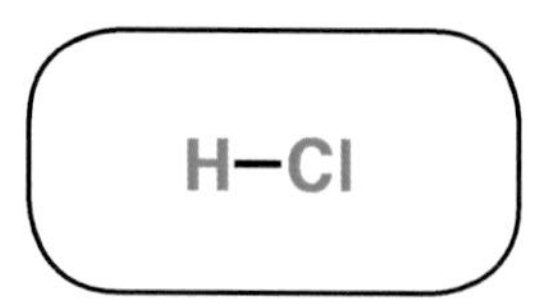

What it's used for: Hydrochloric acid is a strong acid. As a reagent, it can react with multiple bonds in alkenes and alkynes, forming chlorinated compounds. It can also convert alcohols to alkyl chlorides.

Similar to: HBr, HI

Example 1: Hydrohalogenation - conversion of alkenes to alkyl chlorides

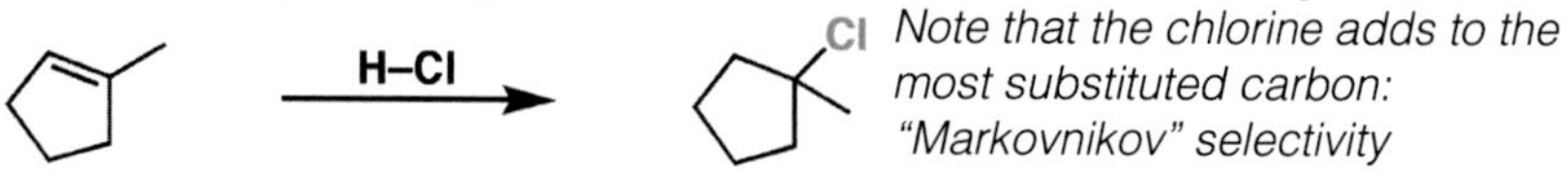

H–Cl
Cl
Note that the chlorine adds to the most substituted carbon: "Markovnikov" selectivity

Example 2: Hydrohalogenation - conversion of alkynes to alkenyl chlorides

H–Cl (1 equiv)

Example 3: Hydrohalogenation - conversion of alkynes to geminal dichlorides

H–Cl (2 equiv)

"geminal" dichloride

Example 4: Conversion of alcohols to alkyl chlorides (S_N2)

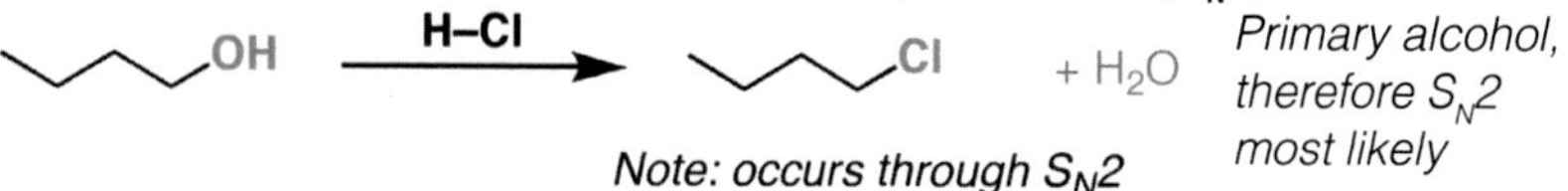

Note: occurs through S_N2

Example 5: Conversion of alcohols to alkyl chlorides (S_N1)

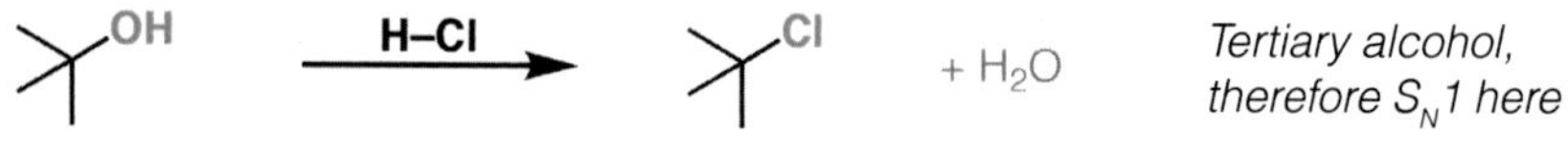

Note: occurs through S_N1

How it works: ***Addition to alkenes***

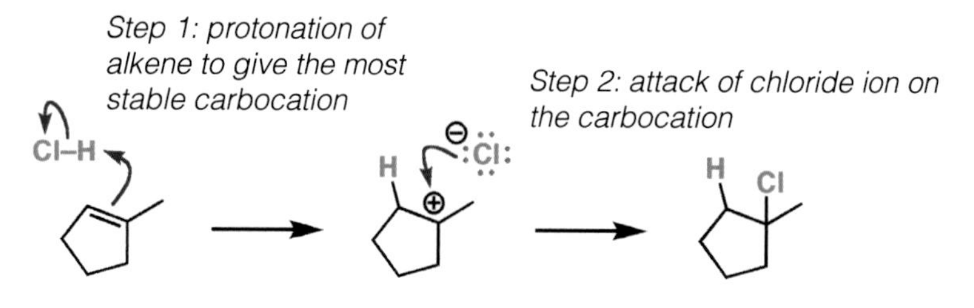

HCl
(continued)

How it works: ***Addition to alkynes***

Addition of 1 equivalent of HBr will lead to a vinyl bromide; addition of a second equivalent leads to the geminal dibromide

Attack of chloride upon carbocation

Formation of most stable carbocation

H–Cl

Attack of chloride upon carbocation

Formation of most stable carbocation

How it works: ***Formation of alkyl chlorides from alcohols***

Protonation of OH by HCl makes a good leaving group (H_2O). When a stable carbocation cannot be formed, the reaction proceeds via an S_N2 pathway:

protonation

backside attack of Cl

+ H_2O

Reaction proceeds via backside attack on the primary carbon

In situations where a more stable carbocation can be formed (e.g. with tertiary alcohols), the reaction proceeds via S_N1:

protonation

loss of leaving group

attack of bromide ion

+ H_2O

H_2CrO_4
Chromic acid

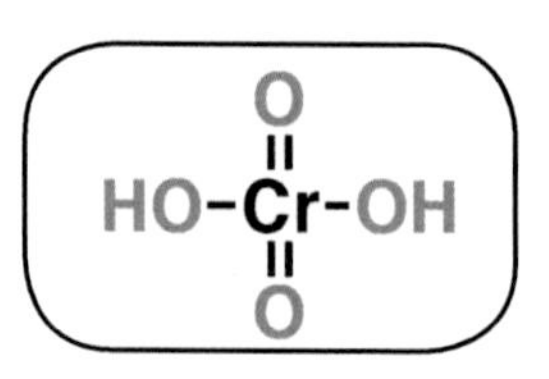

Also known as: Chromic acid is often formed in solution by adding acid to salts of chromate or dichromate. Examples:
$K_2Cr_2O_7$ / H_3O^+, $Na_2Cr_2O_7$ / H_3O^+, $NaCrO_4$/ H_3O^+, $KCrO_4$/H_3O^+, CrO_3/H_3O^+

All of these conditions are equivalent to H_2CrO_4

What it's used for: Chromic acid is a strong oxidizing agent. It will oxidize secondary alcohols to ketones and primary alcohols to carboxylic acids.
Similar to: $KMnO_4$

Example 1: Oxidation of secondary alcohols to give ketones

OH → ($Na_2Cr_2O_7$ / H_2SO_4) → O

Example 2: Oxidation of primary alcohols to give carboxylic acids

OH → ($Na_2Cr_2O_7$ / H_2SO_4) → O, OH

How it works: ***Oxidation of alcohols***
Aqueous acidic conditions convert sodium or potassium dichromate into chromic acid, which is the active oxidant here.

$K^{\oplus}$ $^{\ominus}$O-Cr(=O)(=O)-O-Cr(=O)(=O)-$O^{\ominus}$ $K^{\oplus}$ → (H_2O / H_2SO_4) → 2 HO-Cr(=O)(=O)-OH *Chromic acid*

Chromic acid is then attacked by oxygen. Deprotonation of the C-H bond results in oxidation of the alcohol.

H HO—Cr(=O)(=O)—OH, $H^{\oplus}X^{\ominus}$ → $X^{\ominus}$ H, HO, Cr—OH, OH, H, H, :OH_2 *Water is basic enough to remove a proton here*

$X^{\ominus}$ *is the counterion of the strong acid used e.g. $HSO_4^{\ominus}$ for H_2SO_4*

↓

O, H ← H, O, $X^{\ominus}$, H

+ CrO_2 + H_2O + HX

http://masterorganicchemistry.com

H_2CrO_4 (continued)

How it works: ***Oxidation of aldehydes to carboxylic acids***

$H^{\oplus}X^{\ominus}$ O: H $\xrightarrow{H_2O}$ $X^{\ominus}$ $\oplus$ O H H $H_2\ddot{O}$: $\xrightarrow{deprotonation}$ H HO: O: H HO—Cr—OH O O O $H^{\oplus}X^{\ominus}$

aldehyde hydrate

O HO Cr—OH $X^{\ominus}$ H $\oplus$O OH H OH :OH_2

The base here can be water or the conjugate base of the acid

$\oplus$ O H $X^{\ominus}$ OH

O OH $\xleftarrow{deprotonation}$

+ $H^{\oplus}X^{\ominus}$

+ CrO_2 + H_2O

$Hg(OAc)_2$
Mercuric Acetate

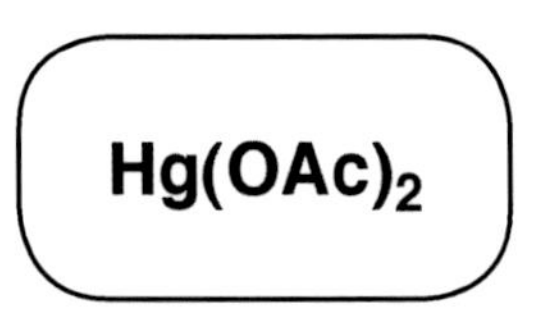

What it's used for: Mercuric acetate is a useful reagent for the oxymercuration of alkenes and alkynes. It makes double bonds more reactive towards nucleophilic attack by nucleophiles such as water and alcohols. The mercury is removed by using $NaBH_4$ (or H_2SO_4 in the case of addition to alkynes).

Similar to: $HgSO_4$

Example 1: Oxymercuration - conversion of alkenes to alcohols

1) $Hg(OAc)_2$, H_2O; 2) $NaBH_4$ → OH + HOAc + Hg

Example 2: Oxymercuration - conversion of alkenes to ethers

1) $Hg(OAc)_2$, $HOCH_3$; 2) $NaBH_4$ → O–CH_3 + HOAc + Hg

Example 3 - Oxymercuration - conversion of alkynes to ketones

—≡—H, $Hg(OAc)_2$, H_2SO_4, H_2O → O + HOAc + Hg

How it works: ***Oxymercuration of alkenes***

In the oxymercuration reaction, an alkene reacts with mercuric acetate to give a 3-membered ring containing mercury (a "mercurinium ion"). This is then attacked by a nucleophilic solvent (e.g. water) at the most substituted carbon

attack at most substituted carbon ("Markovnikov" addition)

Hg, OAc, OAc → :OH_2, ⊕, ⊖OAc, HgOAc → ⊖ :ÖAc, H, ⊕ O–H, HgOAc

mercurinium ion

$NaBH_4$ converts the C–Hg bond to a C–H bond

O–H, HgOAc + HOAc → $NaBH_4$ → OH + BH_3 + NaOAc + Hg (s)

The reduction goes through a free radical on carbon, so there is no syn/anti selectivity

The result of the oxymercuration reaction is a Marvkovnikov addition of water to an alkene. After treatment with $NaBH_4$, solid mercury (0) is obtained.

$Hg(OAc)_2$ (continued)

How it works: ***Oxymercuration of alkynes***

Treatment of an alkyne with $Hg(OAc)_2$ and water leads to the formation of an enol, which converts to a ketone through tautomerization.

Attack of water at the most substituted carbon (Markovnikov addition)

H_2SO_4 *Acid replaces the mercury with H* + HOAc

Tautomerization

Tautomerization favors the ketone

Trivial detail: since mercury is liberated as Hg^{2+}, this process is catalytic in mercury.

$HgSO_4$
Mercuric Sulfate

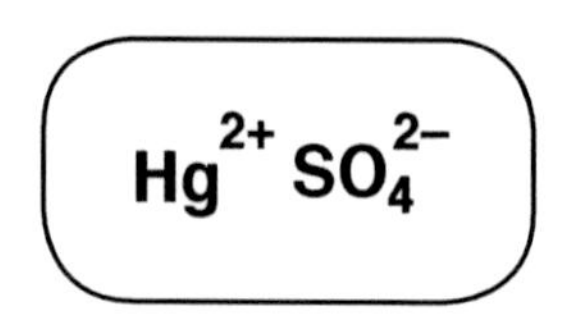

What it's used for: Mercuric sulfate is a Lewis acid. In the presence of aqueous acid ("H_3O^+" or H_2SO_4/H_2O) it will perform the oxymercuration of alkynes to ketones.

Similar to: Mercuric acetate ($Hg(OAc)_2$)

Example 1: Oxymercuration - conversion of alkynes to ketones

$H_3C-≡-CH_3$ → ($HgSO_4$, H_3O^+) → 2-butanone

Phenylacetylene → ($HgSO_4$, H_3O^+) → acetophenone

How it works: ***Oxymercuration of alkynes***

Oxymercuration of alkynes occurs through attack of the alkyne PI bond on Hg^{2+}, followed by attack of water, protonation/demercuration, and tautomerization of the resulting enol to give the ketone.

Formation of mercurinium ion

Attack of H_2O
Note - Markovnikov selective

Deprotonation

H_2SO_4

Attack of enol on H_2SO_4

Loss of mercury

Enol

Tautomerization

HI
Hydroiodic acid

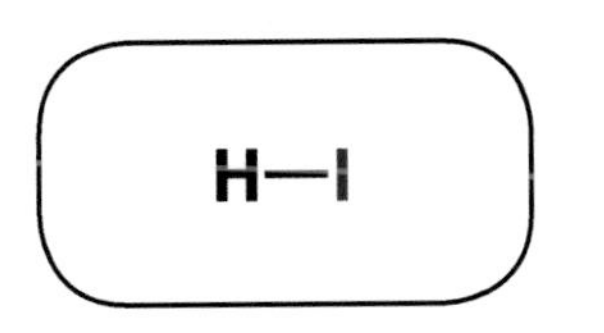

What it's used for: Hydroiodic acid is a strong acid. As a reagent it can add hydrogen and iodine across compounds with multiple bonds such as alkenes and alkynes. It is also useful for the cleavage of ethers and the conversion of alcohols to alkyl halides.
Similar to: HBr, HCl

Example 1: Hydrohalogenation - conversion of alkenes to alkyl iodides

H–I

Note that iodine adds to the most substituted carbon (Markovnikov selectivity)

Example 2: Hydrohalogenation - conversion of alkynes to alkenyl iodides

H–I (1 equiv)

Example 3: Hydrohalogenation - double addition of HI to alkynes to give geminal diiodides

H–I (2 equiv)

"geminal" dihalide

Example 4: Substitution - conversion of alcohols to alkyl iodides (S_N2)

OH → H–I → I + H_2O

Primary alcohol goes through an S_N2 process

Note: occurs through S_N2

Example 5: Substitution - conversion of alcohols to alkyl iodides (S_N1)

OH → H–I → I + H_2O

Tertiary alcohol can form a relatively stable carbocation, therefore an S_N1 process is favorable

Note: occurs through S_N1

Example 6: Conversion of ethers to alcohols and alkyl iodides

O → H–I → OH + I

How it works: ***Addition to alkenes***
In the first step, the alkene is protonated to give the more substituted carbon, followed by attack of the iodide ion to give the alkyl iodide.

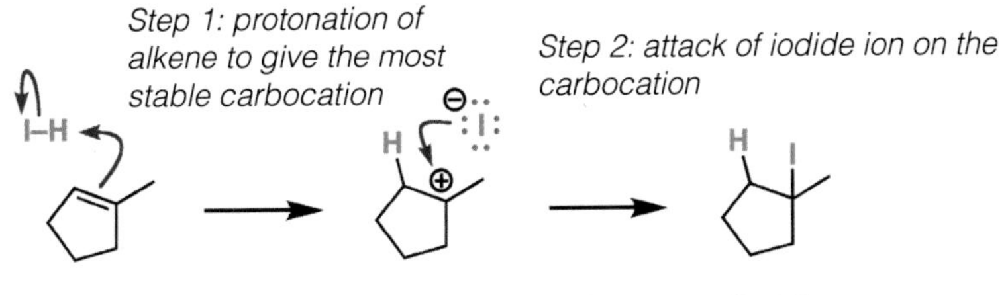

Step 1: protonation of alkene to give the most stable carbocation
Step 2: attack of iodide ion on the carbocation
I–H
H
I
H
I

HI
(continued)

How it works: ***Addition to alkynes***

Addition of 1 equivalent of HI will lead to an alkenyl iodide; addition of a second equivalent leads to the geminal diiodide

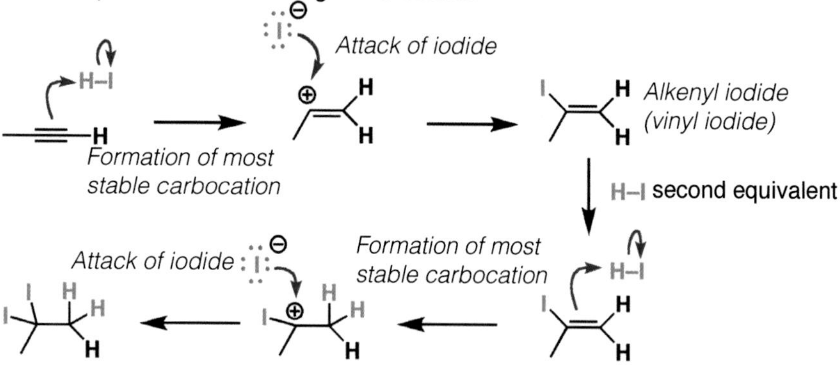

Formation of most stable carbocation

Attack of iodide

Alkenyl iodide (vinyl iodide)

H–I second equivalent

Formation of most stable carbocation

Attack of iodide

How it works: ***Conversion of alcohols to alkyl iodides***

Protonation converts OH to a better leaving group (H_2O). S_N2 dominant for primary

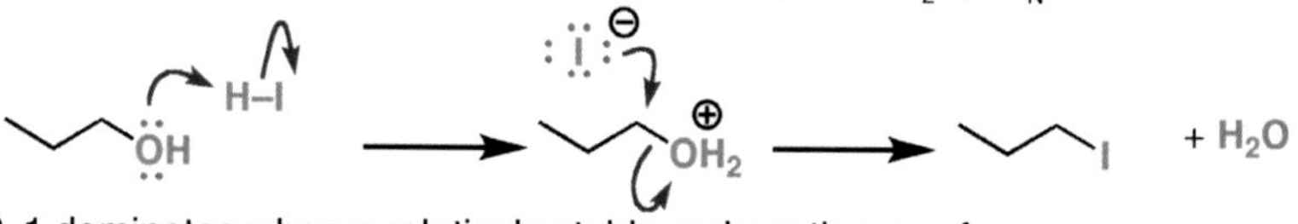

+ H_2O

S_N1 dominates when a relatively stable carbocation can form:

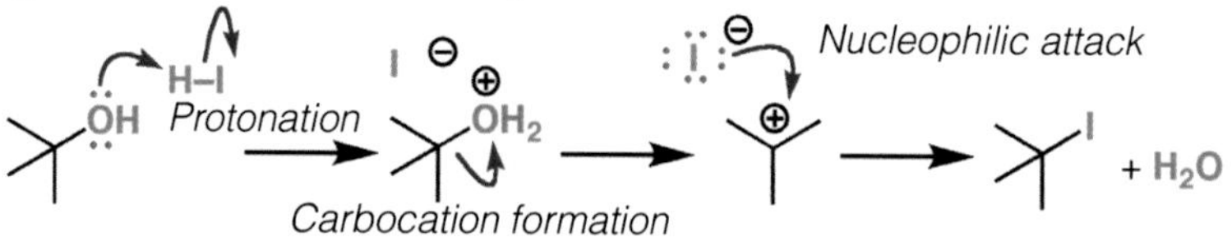

Protonation

Carbocation formation

Nucleophilic attack

+ H_2O

How it works: ***Cleavage of ethers***

Depending on the structure of the ether, cleavage can occur either through S_N1 or S_N2

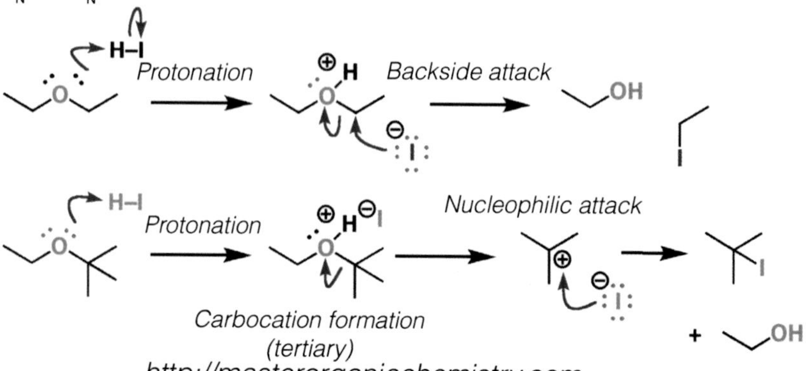

Protonation

Backside attack

Protonation

Carbocation formation (tertiary)

Nucleophilic attack

HIO_4
Periodic acid

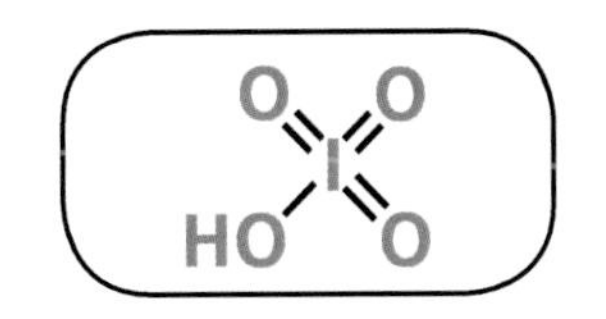

What it's used for: Periodic acid is a strong oxidizing agent. It is most commonly used for the oxidative cleavage of 1,2-diols (vicinal diols) to give aldehydes and ketones.

Similar to: Sodium periodate ($NaIO_4$), Lead (IV) acetate [$Pb(OAc)_4$]

Example 1: Cleavage of diols to give aldehydes/ketones

OH, OH —HIO_4→ O, O, H + HIO_3 + H_2O

OH, OH —HIO_4→ O, H + O, H, H + HIO_3 + H_2O

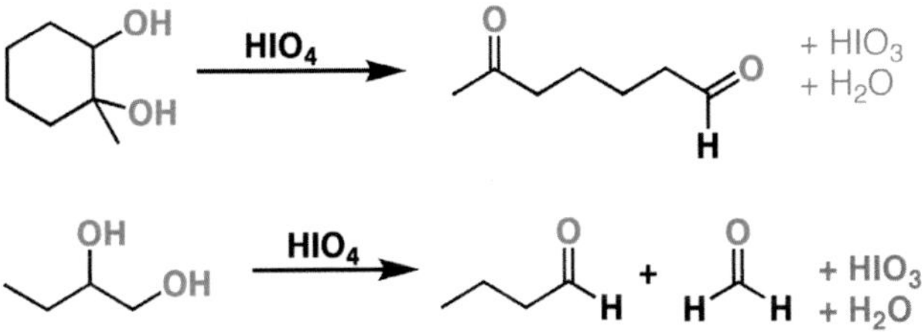

How it works: ***Cleavage of diols to give aldehydes/ketones***

Periodic acid is a strong oxidizing agent. Similar to $Pb(OAc)_4$, it can cleave 1,2-diols (vicinal diols) to give the corresponding aldehydes or ketones.

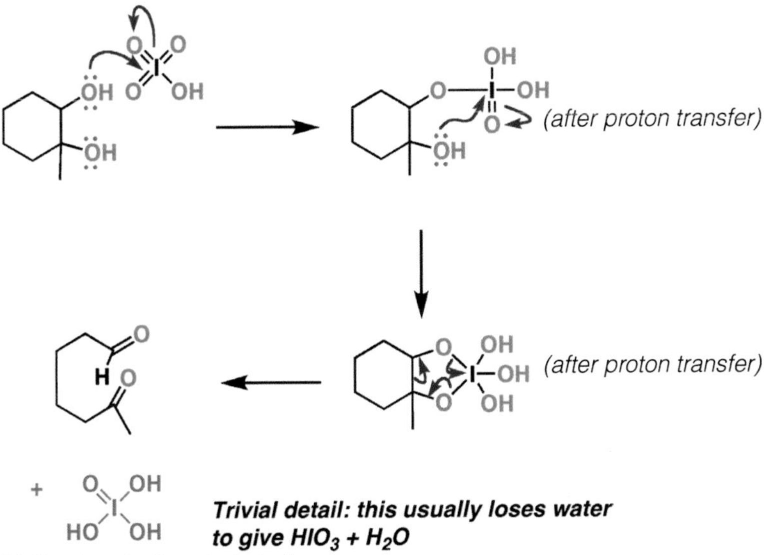

(after proton transfer)

(after proton transfer)

Trivial detail: this usually loses water to give HIO_3 + H_2O

Notice how iodine starts in the (VII) oxidation state and goes to (V) (it has been reduced)

HONO
Nitrous Acid

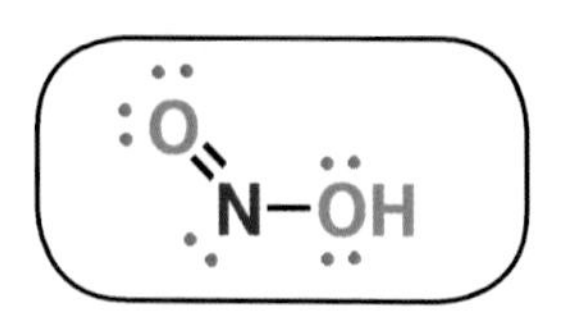

Also known as: HNO_2.

What it's used for: Nitrous acid is primarily used to convert aromatic amines to diazonium salts, which can be converted into many different compounds via the Sandmeyer reaction. It can also be made from $NaNO_2$ if a strong acid such as H_2SO_4 or HCl is added.

Example 1: Conversion of aromatic amines to diazonium salts

$NaNO_2$, HCl

+ H_2O

Diazonium salt

Note: other acids beside HCl can be used as the acid here (such as H_2SO_4)

How it works: ***Formation of diazonium salts***

Nitrous acid reacts with aromatic amines to form diazonium salts. The reaction is greatly assisted by strong acids such as HCl or H_2SO_4

Acid activates the N=O bond toward attack by the amine

HCl

proton transfer

proton transfer

+ H_2O

Diazonium salt

HNO_3
Nitric Acid

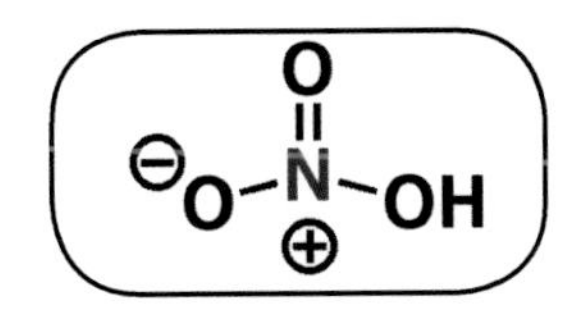

What it's used for: A strong acid, HNO_3 is used as a reagent in the addition of NO_2 to aromatic compounds ("nitration"). It will also oxidize primary alcohols and aldehydes to carboxylic acids.

Example 1: Nitration - conversion of arenes to nitroarenes

HNO_3 / H_2SO_4 → NO_2

H_2SO_4 is a catalyst in this reaction

Example 2: Oxidation - conversion of aldehydes/primary alcohols to carboxylic acids

HNO_3 (dilute) →

This reaction is often introduced in the context of carbohydrate chemistry. Note how the top and botttom carbons have both been oxidized.

How it works: ***Nitration of aromatics***

In the presence of a strong acid such as H_2SO_4, HNO_3 is protonated and loses water to form the nitronium ion (NO_2^+), a very reactive electrophile.

$H\text{–}OSO_3H$ — *Protonation* — HSO_4^- — *Loss of water* — HSO_4^- + H_2O

Nitronium ion

The NO_2^+ is then attacked by the aromatic ring to give a carbocation, which is then deprotonated to restore aromaticity. H_2SO_4 is regenerated which can then go on to react with another equivalent of HNO_3 (it is a catalyst here).

Step 1: Electrophilic addition

HSO_4^- → $^-OSO_3H$ → *Step 2: Deprotonation*

$HOSO_3H$ (H_2SO_4)

Note that acid is regenerated here; it acts as a catalyst in this reaction.

H_2O_2
Hydrogen peroxide

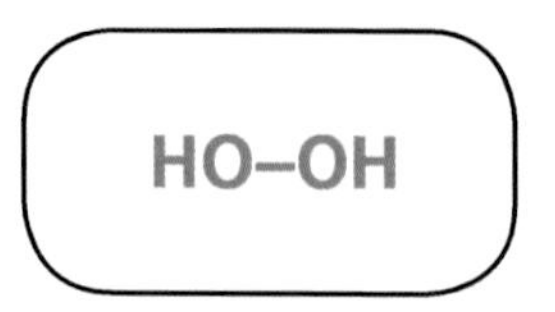

What it's used for: Hydrogen peroxide is used as an oxidant in the hydroboration of alkenes and alkynes, converting the C–B bond into a C–O bond. It is also used in the oxidative workup of ozonolysis, converting aldehydes into carboxylic acids.

Example 1: As an oxidant in the hydroboration reaction

1) BH_3
2) NaOH, H_2O_2

OH

Example 2: For oxidative workup in ozonolysis

1) O_3
2) H_2O_2

O OH + O OH

Note that carboxylic acids (not aldehydes) are the products here

How it works: ***Hydroboration of alkenes***

Hydrogen peroxide is used in concert with the strong base NaOH. Deprotonation of H_2O_2 gives its conjugate base, which is a more reactive nucleophile:

HO⊖ ⊕Na + H–O–O–H ⇌ HO–H + ⊕Na ⊖O–O–H

The peroxide ion then attacks boron. In the key step a rearrangement occurs, breaking the weak (138 kJ/mol) O–O bond.

Attack of peroxide → *Rearrangement* → *Hydrolysis*

The B–O bond is then broken through further attack of NaOH upon boron. The negatively charged oxygen ("alkoxide") is eventually protonated by water.

H–OH; + NaOH; + BH_2OH

H_2O_2
(continued)

How it works: ***Oxidative workup for ozonolysis***

Hydrogen peroxide can oxidize aldehydes to carboxylic acids. This is ithe basis for "oxidative workup" of the ozonolysis reaction, where the aldehyde is treated with aqueous H_2O_2 (minor note: addition of a base such as NaOH speeds up this reaction)

H_2O_2

H_2O

Ozonide (breaks open upon heating)

Addition of H_2O_2 to aldehyde

Proton transfer

Deprotonation

Carboxylic acid

Note: the final deprotonation step is the rate-limiting step; it is greatly accelerated by addition of base.

H_3PO_4
Phosphoric acid

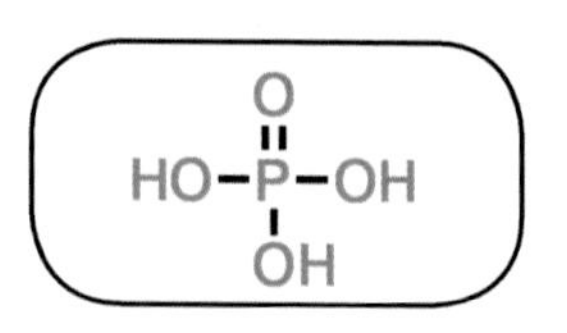

What it's used for: Phosphoric acid is a moderately strong acid. The conjugate acid of H_3PO_4 is a poor nucleophile, so phosphoric acid is an excellent acid to use for elimination reactons.

Similar to: Sulfuric acid (H_2SO_4), tosic acid (TsOH)

Example 1: Elimination of alcohols to give alkenes

OH — H_3PO_4, Δ (heat) → + H_3PO_4 + H_2O

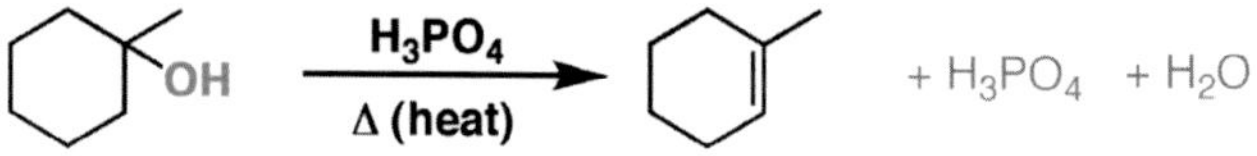

How it works: *Elimination of alcohols to give alkenes*

Protonation of the alcohol by phosphoric acid makes OH into a good leaving group (H_2O) which departs to give a carbocation. Deprotonation of the carbon adjacent to the carbocation leads to an alkene (this is an E1 mechanism).

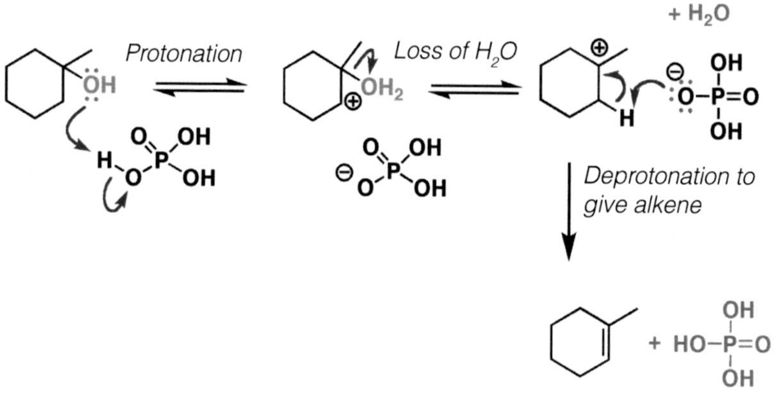

Protonation

Loss of H_2O

+ H_2O

Deprotonation to give alkene

+ HO–P=O (OH)(OH)

H_2SO_4
Sulfuric acid

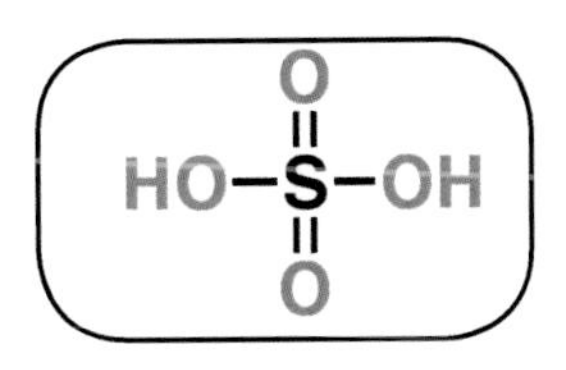

What it's used for: Sulfuric acid, known to alchemists as "oil of vitriol" is a strong acid (pKa –3.0). It is particularly useful for elimination reactions since its conjugate base [HSO_4^-] is a very poor nucleophile. It finds use in many other reactions as a general strong acid.

Similar to: *p*-toluenesulfonic acid (TsOH)

Example 1: Elimination – conversion of alcohols to alkenes

OH —H_2SO_4, Δ (heat)→ + H_2SO_4 + H_2O

How it works: ***Elimination of alcohols***

Protonation of the alcohol forms its conjugate acid [an "oxonium ion"], which has a much better leaving group (H_2O) than the alcohol (HO^-). Loss of water results in the formation of a carbocation. The resonance stabilized HSO_4 anion is a poor nucleophile, and tends not to add to the carbocation (unlike HBr and HCl for example). Deprotonation, either by HSO_4^- or by water, leads to formation of the alkene and regeneration of acid.

Protonation · *Carbocation formation* · + H_2O · *Deprotonation*

Note resonance stability of the HSO_4^- anion

H_2SO_4 is regenerated (i.e. it is a catalyst)

As with many reactions that pass through carbocations, rearrangements can occur in situations where a more stable carbocation can form through a hydride or alkyl shift.

I_2
Iodine

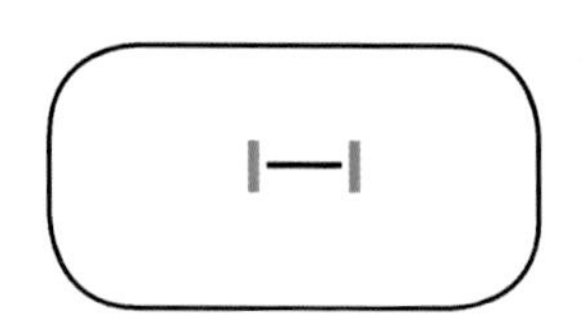

What it's used for: Iodine is an excellent electrophile due to the weak I–I bond (approx 151 kJ/mol [36 kcal/mol]). It reacts with carbon-carbon multiple bonds such as alkenes and alkynes, along with other nucleophiles. It is also used in the iodoform reaction.

Similar to: *N*-iodo succinimide (NIS) performs many of the same reactions.

Example 1: Iodination - conversion of alkenes to vicinal diiodides

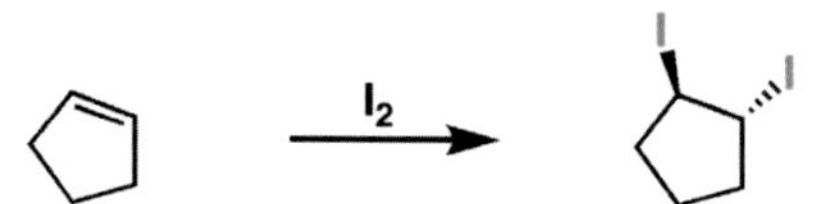

Example 2: Conversion of alkenes to iodohydrins

I_2, H_2O → + HI

Example 3: Conversion of ketones to α-iodo ketones

I_2, $H^{\oplus}$ → + H–I

H^+ here refers to any generic strong acid

Example 4: Iodoform reaction - conversion of methyl ketones to carboxylic acids

NaOH, I_2 → + HCI_3 + NaI

How it works: ***Iodination of alkenes***

Treatment of an alkene with I_2 leads to formation of an iodonium ion, which undergoes backside attack by iodide ion to give the trans product.

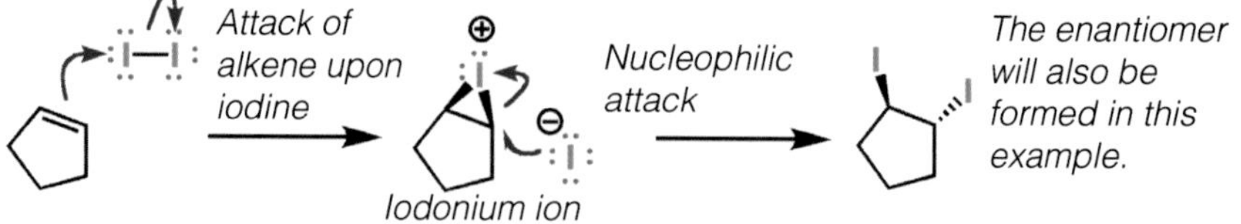

When a nucleophilic solvent such as H_2O or an alcohol is present, halohydrins can form:

I_2, H_2O

A halohydrin

$H_3O^{\oplus}$ $I^{\ominus}$

Iodine
(continued)

How it works: ***Iodination of ketones***

Treatment of ketones with acid ["HX"] catalyzes keto-enol tautomerization. Attack of iodine by the enol tautomer followed by deprotonation gives the iodinated ketone.

H⊕ X⊖
I_2
tautomerization
+ HI

How it works: ***The haloform reaction***

Methyl ketones treated with strong base (e.g. NaOH) form enolates, which attack iodine. After complete replacement of H by I, the $^{-}CI_3$ ion can then be displaced from the ketone, giving a carboxylic acid.

Enolate formation
Iodination
+ NaI
Enolate formation
Iodination
Enolate formation
Iodination
Addition of ^{-}OH
Elimination of the $^{-}CI_3$ ion (a weak base)
+ HCI_3

Extra detail: under the basic reaction conditions, the carboxylic acid will be deprotonated. Acidic workup required.

$KMnO_4$
Potassium permanganate

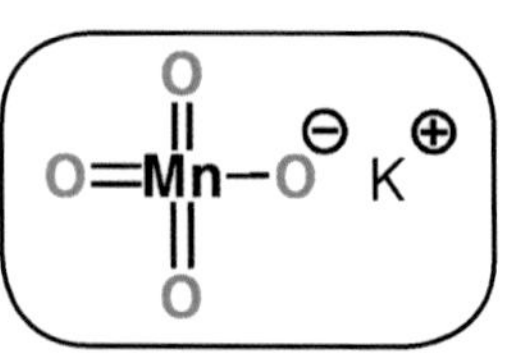
O
O=Mn–O⊖ K⊕
O

What it's used for: This strong oxidizing agent will oxidize primary alcohols (and aldehydes) to carboxylic acids, secondary alcohols to ketones, form diols from alkenes, and oxidatively cleave carbon-carbon bonds. It will also oxidize C-H bonds adjacent to aromatic rings.

Similar to: $K_2Cr_2O_7$, OsO_4, O_3

Example 1: Oxidation - conversion of primary alcohols to carboxylic acids

OH —$KMnO_4$→ O, OH

Example 2: Oxidation - conversion of secondary alcohols to ketones

OH —$KMnO_4$→ O

Example 3: Oxidation - conversion of aldehydes to carboxylic acids

O, H —$KMnO_4$→ O, OH

Example 4: Oxidative cleavage - conversion of alkenes to ketones / carboxylic acids

—$KMnO_4$, $H_3O^{\oplus}$→ OH, O + O, OH

—$KMnO_4$, $H_3O^{\oplus}$→ OH, O + O

note: ketone

Example 5: Dihydroxylation - conversion of alkenes to vicinal diols

—$KMnO_4$, cold, dilute NaOH→ OH, OH

Note that the stereochemistry of the diol is "syn"

Example 6: Oxidation - conversion of alkylaromatics to carboxylic acids

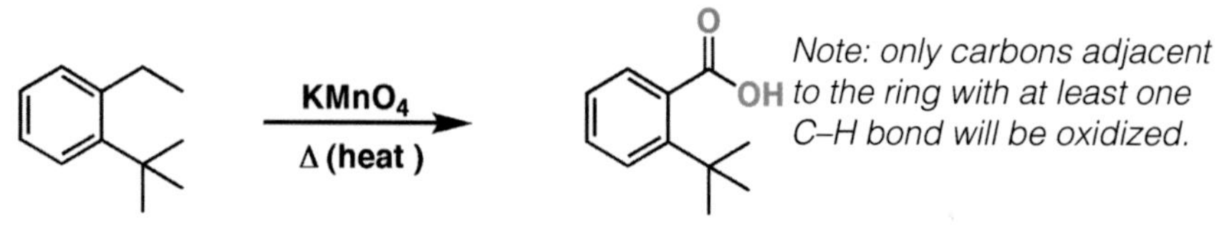
KMnO4
Δ (heat)
O
OH
Note: only carbons adjacent to the ring with at least one C–H bond will be oxidized.

$KMnO_4$ (continued)

How it works: ***Oxidation of primary and secondary alcohols***
Primary alcohols are oxidized to carboxylic acids; secondary alcohols are oxidized to ketones.

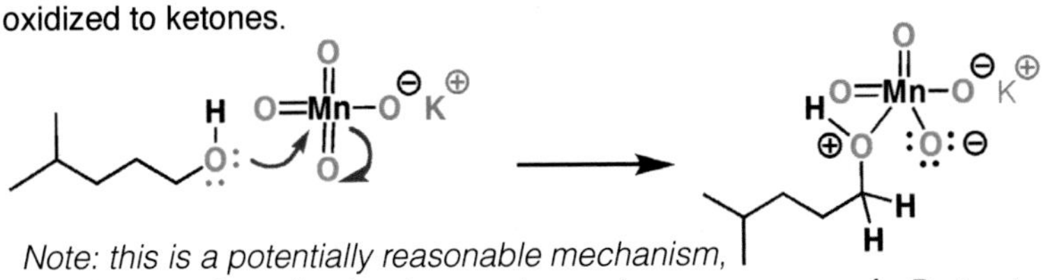

Proton transfer

Note: this is a potentially reasonable mechanism, but the actual mechanism is complex and can go down different pathways (some involving free radicals, and some not fully understood!) Take this with a grain of salt.

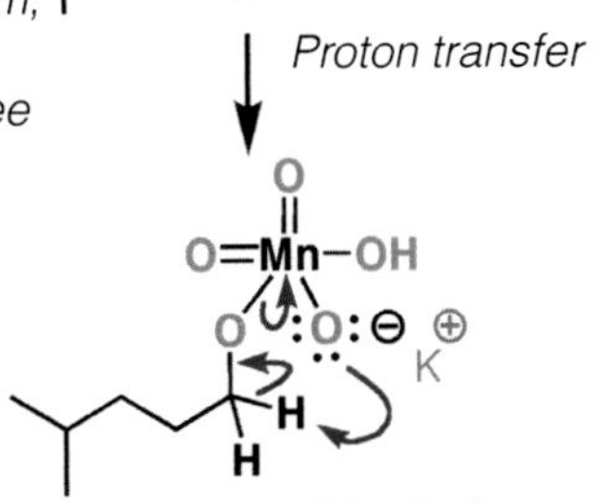

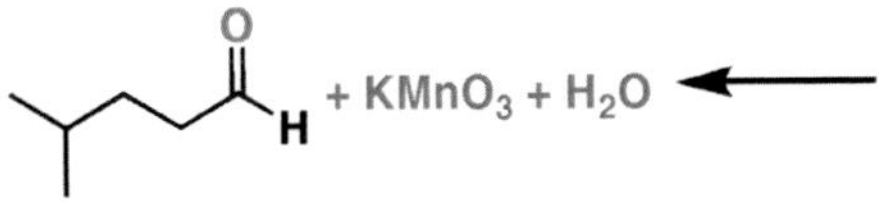

+ $KMnO_3$ + H_2O

Aldehydes are not stable under these reaction conditions and will be further oxidized to carboxylic acids

H_2O

Aldehyde hydrate

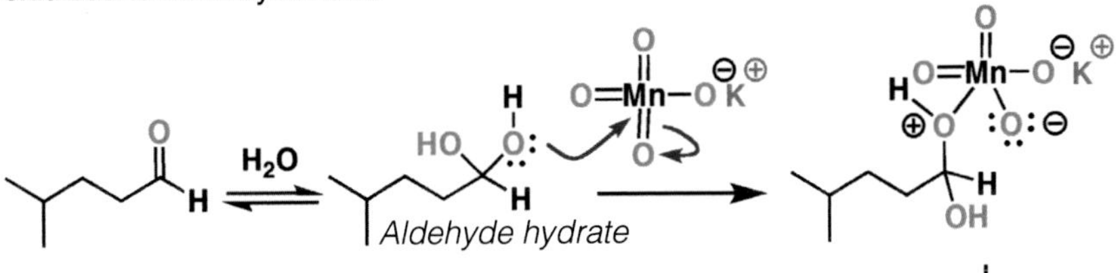

Again, a suggested mechanism here, but $KMnO_4$ oxidation can occur via multiple pathways (beyond the scope of our discussion).

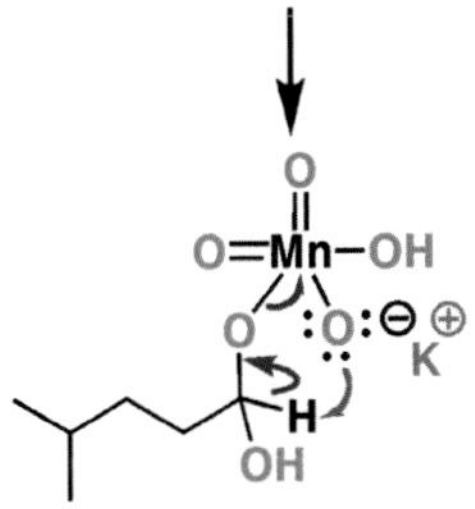

+ $KMnO_3$ + H_2O

How it works: ***Oxidation of aromatic side chains***
The mechanism for side chain oxidation is complex (involves free radicals) and not completely understood. It requires that the carbon adjacent to the arene have at least one C–H bond.

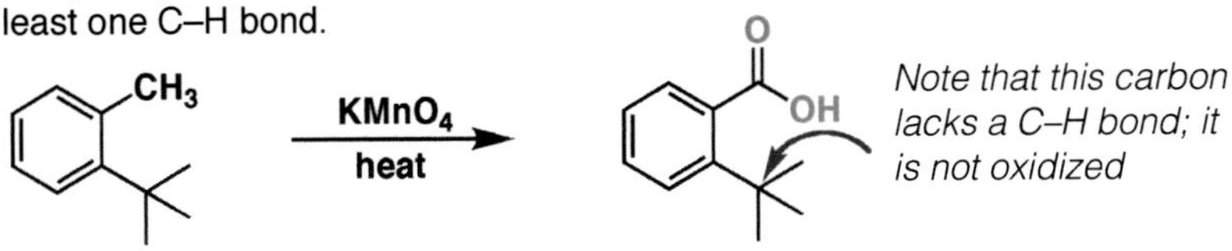

$KMnO_4$
heat

Note that this carbon lacks a C–H bond; it is not oxidized

$KMnO_4$ (continued)

How it works: ***Dihydroxylation of alkenes***

Under cold, dilute basic conditions, $KMnO_4$ will convert alkenes into 1,2-diols (vicinal diols). Yields for this process are typically lower than for OsO_4.

KOH

In the absence of base, diols undergo oxidative cleavage (see below)

Base (KOH) cleaves the cyclic Mn compound ("manganate ester").

+ $KMnO_3$

proton transfer

$H^{\oplus}$ *Acid workup leads to the diol*

How it works: ***Oxidative cleavage of alkenes***

Under acidic conditions vicinal diols undergo oxidative cleavage.

$KMnO_4$ is protonated to give $HMnO_4$ and adds to the alkene as above:

Addition

Oxidative cleavage

+ MnO_2

"cyclic manganate ester"

Oxidation to carboxylic acid (as above)

KO*t*-Bu
Potassium *t*-butoxide

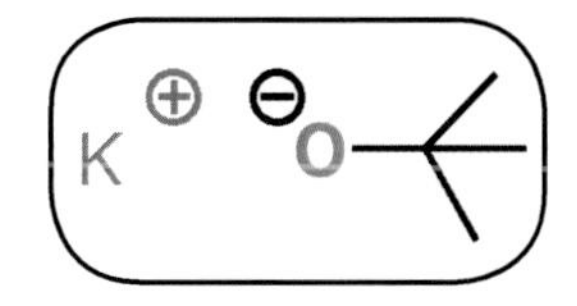

Also known as: $KOC(CH_3)_3$, potassium tert-butoxide

What it's used for: Potassium *t*-butoxide is a strong, sterically hindered base. The prototypical "bulky base", it is useful in elimination reactions for forming the less substituted "non-Zaitsev" [sometimes called "Hofmann"] alkene product.

Similar to: Essentially identical to NaOtBu and LiOtBu [these are all treated as the same here]. Lithium diisopropyl amide (LDA) is a stronger bulky base.

Example 1: Elimination - conversion of alkyl halides to alkenes ("non-Zaitsev" or "Hofmann" alkene products)

Br **$KOC(CH_3)_3$** → +

Hofmann product (major)

Zaitsev product (minor)

+ KBr + $HOC(CH_3)_3$

How it works: ***Formation of "non-Zaitsev" elimination products***

Elimination reactions generally favor formation of the more substituted alkene ("Zaitsev's rule"). However, steric clash between the bulky base and alkyl groups can disfavor this pathway.

H Br

K O H H H Br

"non-Zaitsev" or "Hofmann" pathway

Favored

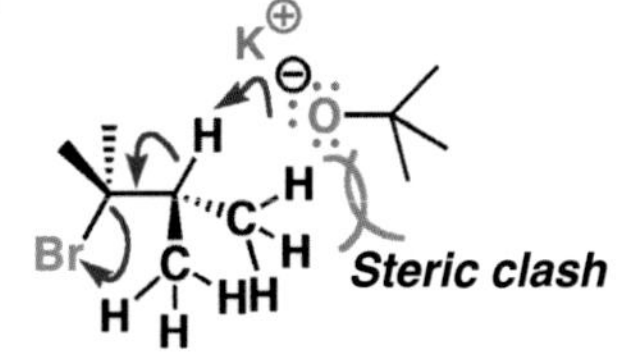

"Zaitsev" pathway

Disfavored

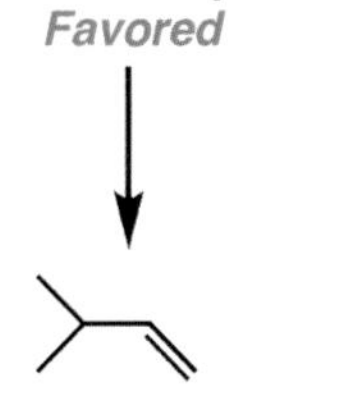

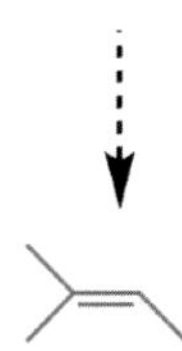

The less substituted alkene ("non-Zaitsev product") is the major product here

The more substituted alkene ("Zaitsev product") is the minor product here

KPhth
Potassium Phthalimide

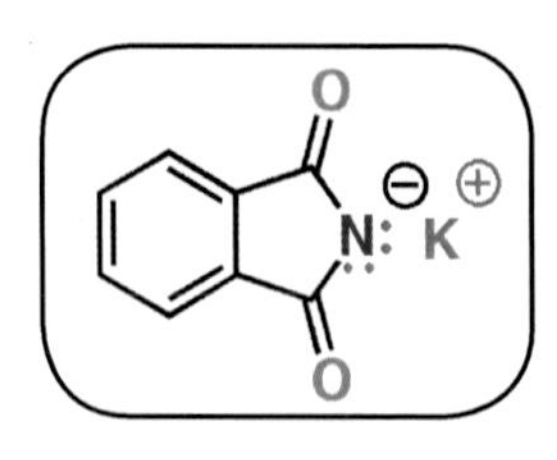

Also known as: phthalimide ion

What it's used for: Sodium (or potassium) phthalimide is a nitrogen-containing nucleophile used in the Gabirel synthesis. Potassium phthalimide reacts with alkyl halides to form a C–N bond, which is then cleaved by treatment with hydrazine to give a primary amine.

Example 1: Substitution - formation of alkyl phthalimides from alkyl halides

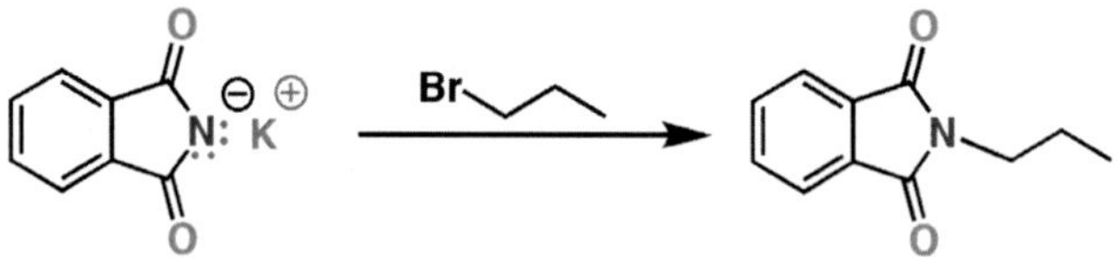

Example 2: Conversion of phthalimides to primary amines (after cleavage with NH_2NH_2)

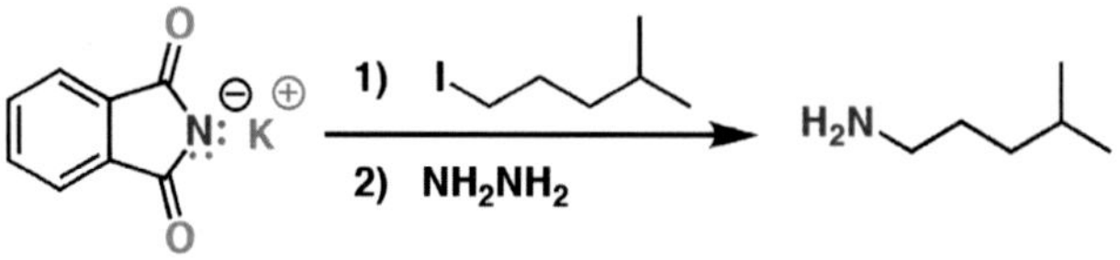

How it works: ***Substitution reaction***

In the reaction below, a strong base (NaH) is used to deprotonate phthalamide to give the conjugate base, which then performs an S_N2 reaction on a primary alkyl halide.

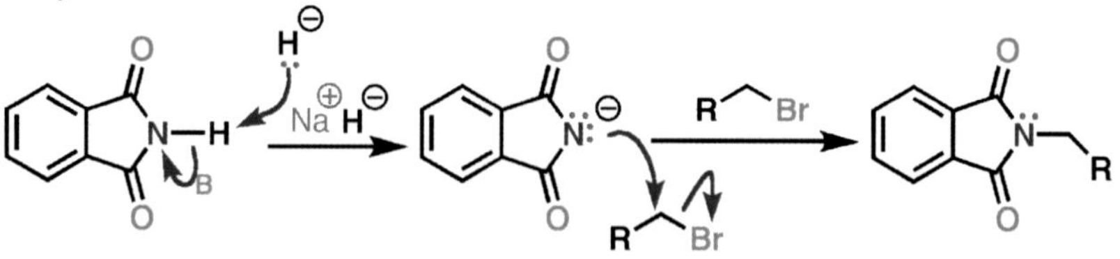

LDA

Lithium diisopropyl amide

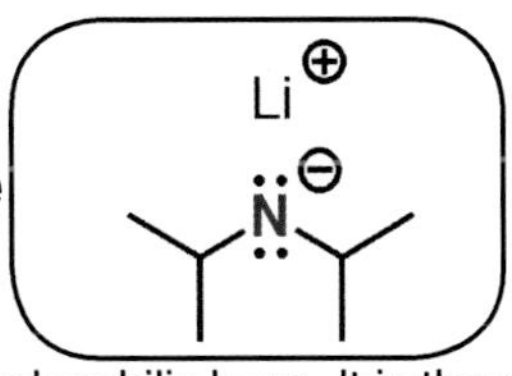

What it's used for: LDA is a strong, bulky, non-nucleophilic base. It is the reagent of choice for selectively removing a proton from the least hindered carbon next to a ketone. It can also be used to form the "Hofmann" product in elimination reactions.

Similar to: $NaNH_2$ (in strength), KOt-Bu (in size)

Example 1: Conversion of ketones to enolates

O → LDA, THF, –78°C → OLi + H–N (Diisopropyl amine)

Note that deprotonation occurs at the least substituted carbon

Diisopropyl amine

Example 2: Elimination of alkyl halides to give "Hofmann" alkenes

Br → LDA →

"Hofmann" alkene product (major)

"Zaitsev" product (minor)

How it works: ***Formation of less substituted enolates ("kinetic" enolates)***

The bulky isopropyl groups of LDA make it a highly selective base for removing a proton from the less hindered α-carbon of the ketone.

O → LDA, THF, –78°C → Li⊕ ⊖N, H, H, H, O

the isopropyl group is bulkier than the methyl group

Tetrahydrofuran (THF) is a common solvent for this reaction. Low temperature maximizes selectivity.

[Li⊕ ⊖H_2C O ⟷ H, H, :Ö: ⊖ Li ⊕] + H–N

(Resonance forms of the enolate)

Diisopropylamine (the conjugate acid of LDA)

Li
Lithium

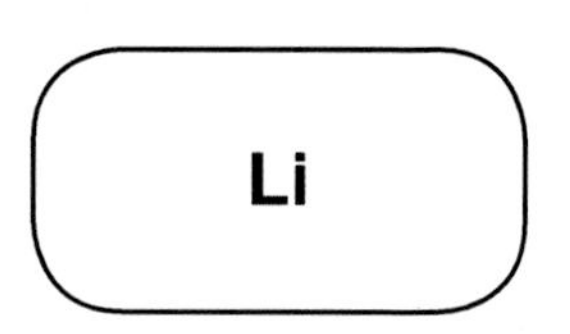

What it's used for: Lithium is a reducing agent. It will convert alkyl halides to alkyl lithium compounds. It is similar to (although a weaker reducing agent than) sodium and potassium. It will also form H_2 when treated with alcohols, giving lithium alkoxides.

Similar to: Sodium (Na), Potassium (K)

Example 1: Conversion of alkyl halides to alkyllithiums

Br → Li (2 equiv) → Li + LiBr

Example 2: Conversion of alcohols to alkoxides

OH → Li → OLi + 1/2 H_2

Example 3: Birch reduction - conversion of arenes to dienes

O → Li, NH_3, t-BuOH → O + t-BuOLi

How it works: ***Formation of organolithium reagents***

Like all alkali metals, lithium readily gives up its single valence electron. When treated with an alkyl halide, it will form an alkyl lithium species. Two equivalents of lithium are required for this reaction.

Br •Li → Br ⊖ ⊕ Li

This is called a "radical anion"

•Li + LiBr

Li

Alkyllithiums: strong bases and excellent nucleophiles.

Li
(continued)

How it works: ***Birch reduction***

The Birch reduction is a useful way of obtaining dienes from aromatic groups. Ammonia (NH_3) is the usual solvent with small amounts of an alcohol such as *t*-BuOH providing a source of protons

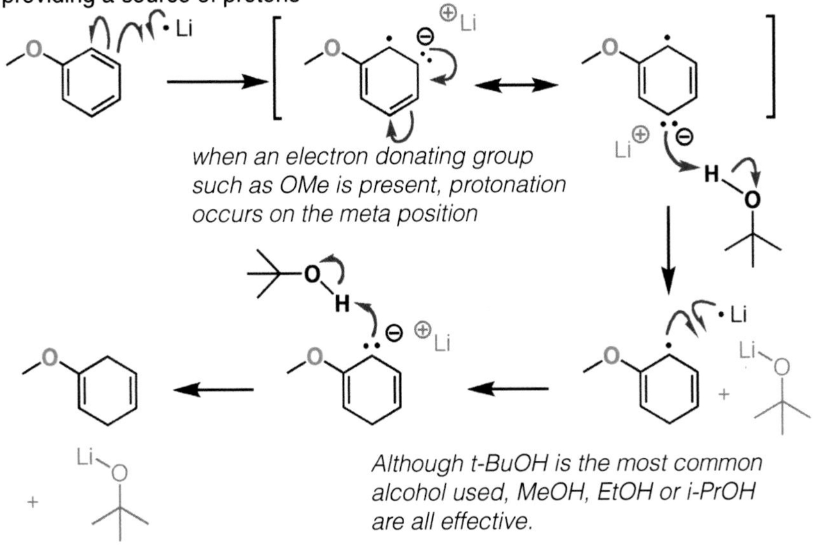

when an electron donating group such as OMe is present, protonation occurs on the meta position

Although t-BuOH is the most common alcohol used, MeOH, EtOH or i-PrOH are all effective.

When electron withdrawing substituents are present, protonation occurs on the carbon adjacent to the electron withdrawing group

Note that protonation occurs adjacent to the electron withdrawing group

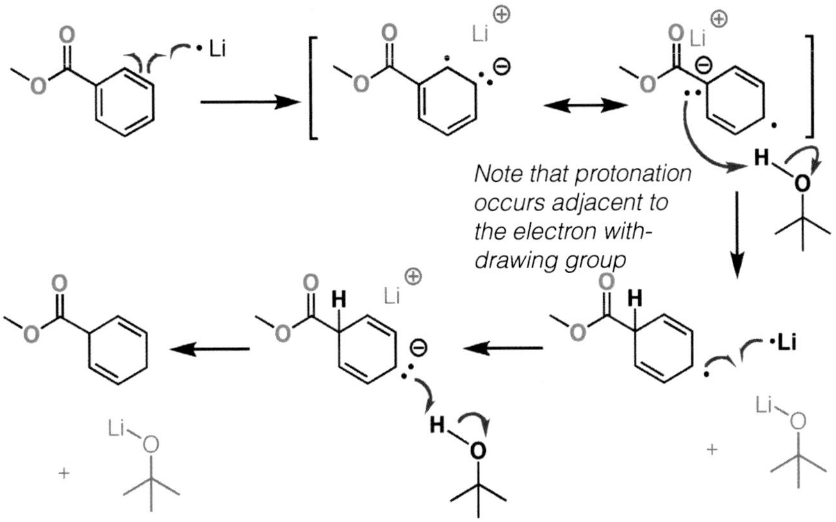

Lindlar's Catalyst

$Pd\text{-}CaCO_3\text{-}PbO_2$

Also known as: Poisoned catalyst, $Pd\text{-}CaCO_3$

What it's used for: Lindlar's catalyst is a poisoned palladium metal catalyst that performs partial hydrogenations of alkynes in the presence of hydrogen gas (H_2). It always gives the *cis* alkene, in contrast to Na/NH_3 which gives *trans*.
Similar to: Nickel boride (Ni_2B), palladium on barium sulfate, $Pd\text{-}CaCO_3$-quinoline

Example 1: Lindlar reduction - conversion of alkynes to alkenes

Lindlar / H_2

Lindlar / H_2

How it works: ***Partial hydrogenation***
Other than its lower activity when compared with non-poisoned metal catalysts, Lindlar's catalyst behaves in all ways similar to other heterogeneous metal catalysts such as Pd/C, Pt, Ni, etc. (see these seperately). The alkyne and hydrogen are adsorbed on to the metal surface and delivered in *cis* fashion.

Sometimes the aromatic amine quinoline is used, which assists the selectivity of the reaction and prevents the formation of alkanes.

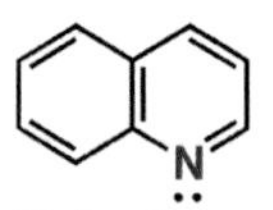

Quinoline

It is thought that the role of Pb (lead) is to reduce the amount of H_2 adsorbed, while quinoline helps to prevent the formation of unwanted byproducts.

$LiAlH_4$
Lithium aluminum hydride

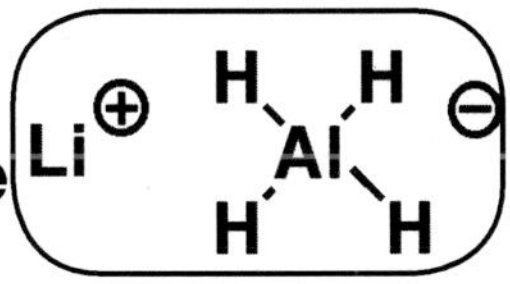

Also known as: LAH

What it's used for: Lithium aluminum hydride is a very strong reducing agent. It will reduce aldehydes, ketones, esters and carboxylic acids to alcohols, amides and nitriles to amines, and open epoxides to give alcohols.

Similar to: $NaBH_4$, DIBAL, $LiAlH(Ot\text{-}Bu)_3$

Example 1: Reduction - conversion of esters to primary alcohols

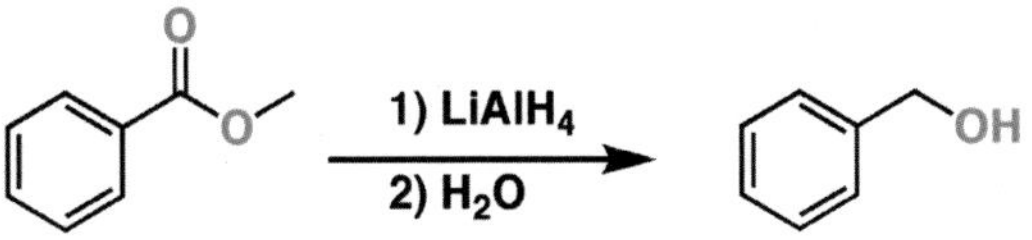

1) $LiAlH_4$
2) H_2O

Example 2: Reduction - conversion of carboxylic acids to primary alcohols

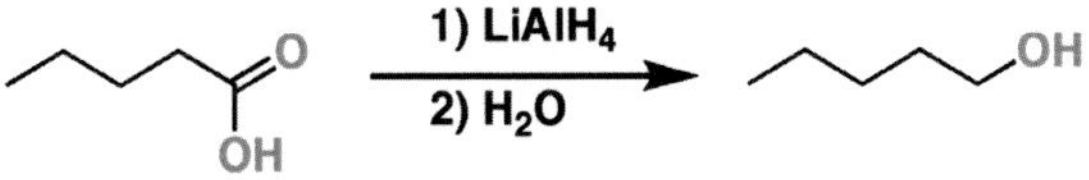

1) $LiAlH_4$
2) H_2O

Example 3: Reduction - conversion of amides to primary amines

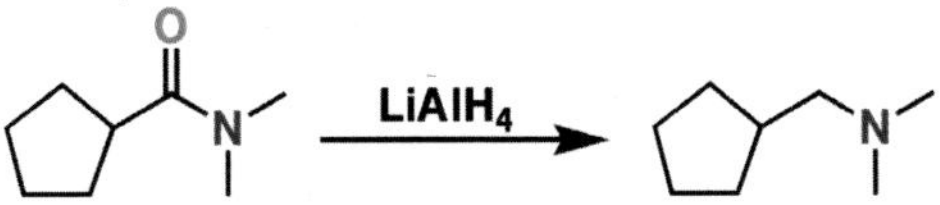

$LiAlH_4$

Example 4 - Reduction - conversion of ketones to secondary alcohols

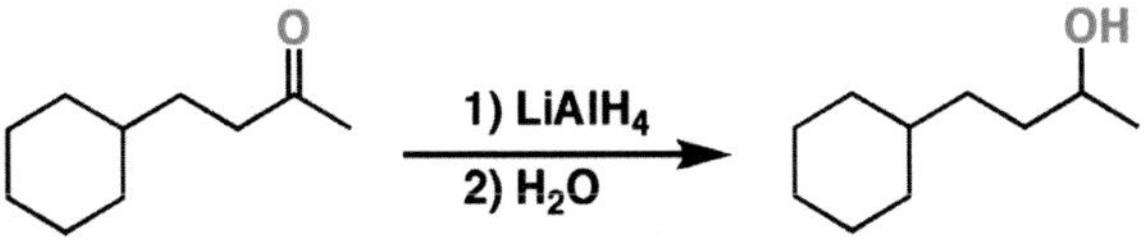

1) $LiAlH_4$
2) H_2O

Example 5 - Reduction - conversion of nitriles to primary amines

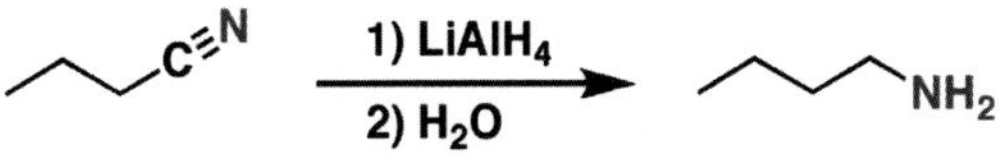

1) $LiAlH_4$
2) H_2O

Example 6 - Reduction - conversion of azides to primary amines

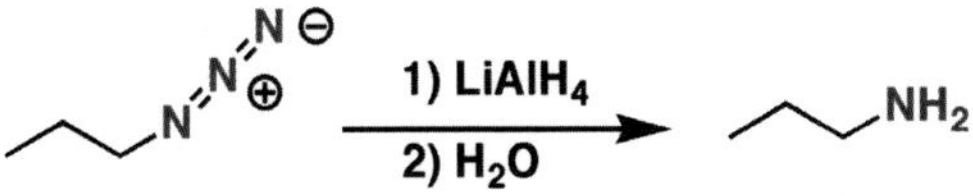

1) $LiAlH_4$
2) H_2O

Example 7 - Reduction - conversion of epoxides to alcohols (ring opening)

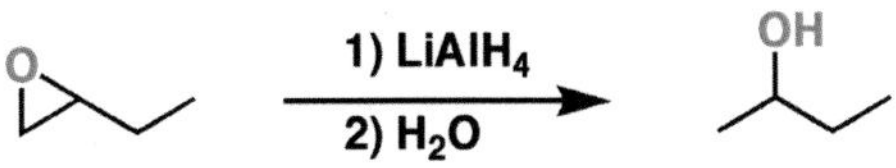

1) $LiAlH_4$
2) H_2O

$LiAlH_4$ (continued)

How it works: ***Reduction of esters, amides, and nitriles***
Lithium aluminum hydride is a very strong reducing agent capable of reacting with a wide variety of functional groups. It is generally not possible to control reactions of $LiAlH_4$ so that they "stop" part of the way; reactions of esters go straight to alcohols, for instance.

Reduction of esters:

AlH_3 Li · OCH3 · H · + $LiOCH_3$ · H · Al · H · H

The reaction does not stop at the aldehyde stage, going straight through to the alcohol

OH · H · H · H_2O · O · AlH_2 · H · H

Reduction of amides:

NMe2 · H · AlH_3 Li · AlH_3 Li · O · H · NMe2 · Li · O · H_2Al—H · N · Iminium ion

Here, oxygen is a better leaving group than nitrogen!

N · + H_2AlO Li

Reduction of nitriles:

C≡N · H · AlH_3 Li · N Li · H · H_2Al—H · N Li

NH_2 · H_2O (workup) · AlH_2 · N Li

$LiAlH(Ot\text{-}Bu)_3$

Lithium tri tert-butoxy aluminum hydride

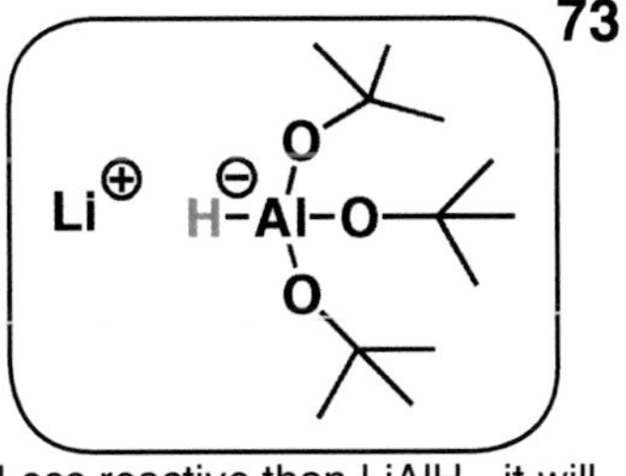

Also known as: $LiAlH[OC(CH_3)_3]$

What it's used for: Strong, bulky reducing agent.Less reactive than $LiAlH_4$, it will convert acyl halides to aldehydes.

Similar to: $NaBH_4$, DIBAL, $LiAlH_4$

Example 1: Reduction of acyl halides to aldehydes

O Cl —$LiAlH[OC(CH_3)_3]_3$→ O H

Acyl bromides will react similarly

How it works: ***Reduction of acyl chlorides***

The mechanism for this reaction is similar to $LiAlH_4$. So long as only one equivalent is used, the aldehyde will not be reduced further to the alcohol.

O Cl Li⊕ H—Al—O ⊖O O → O⊖ Li⊕ Cl H → O H + LiCl + $Al[OC(CH_3)_3]$

m-CPBA

m-chloroperoxybenzoic acid

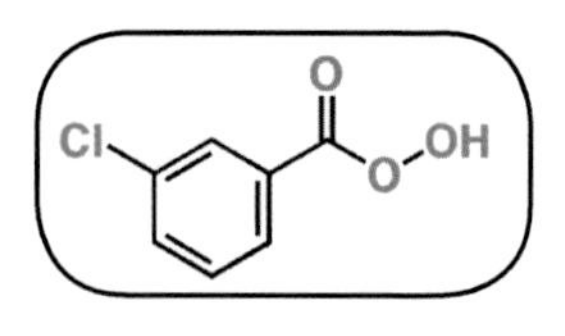

What it's used for: mCPBA (sometimes written MCPBA) is an oxidizing agent. It sees use in two main ways. First, it is used to transform alkenes into epoxides. Secondly, it will react with ketones to form esters in the Baeyer-Villiger reaction.

Similar to: Peroxyacetic acid [CH_3CO_3H], trifluoroperoxyacetic acid [CF_3CO_3H]

Example 1: Epoxidation - conversion of alkenes to epoxides

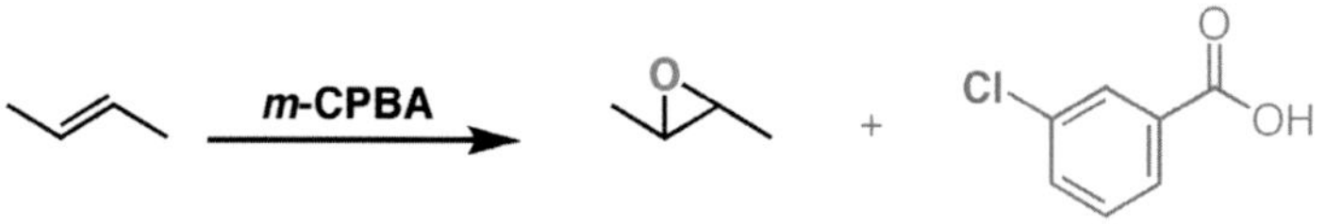

m-CPBA

+

m-chlorobenzoic acid (byproduct)

Example 2: Baeyer-Villiger reaction - conversion of ketones to esters

m-CPBA

How it works: ***Epoxidation of alkenes***

Treatment of alkenes with mCPBA leads to the formation of epoxides through a concerted transition state. The reaction is completely stereospecific: trans alkenes and cis alkenes give stereoisomeric products.

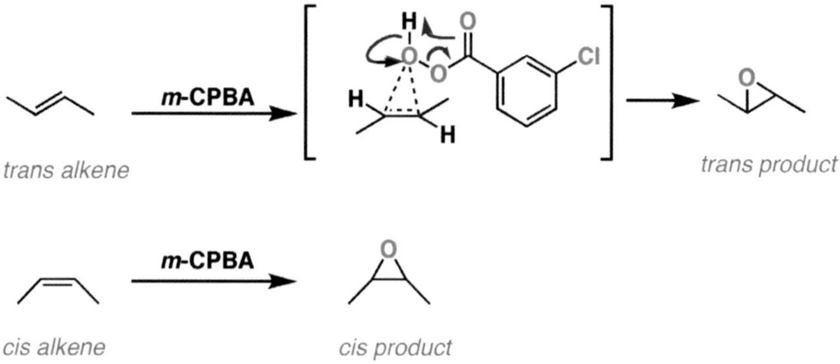

m-CPBA

trans alkene

trans product

m-CPBA

cis alkene

cis product

m-CPBA
(continued)

How it works: ***Baeyer-Villiger reaction***

In the Baeyer-Villiger reaction, m-CPBA adds to a ketone to form a tetrahedral intermediate. In the key step, the carbon migrates to oxygen, breaking the weak O–O bond and leading to the formation of an ester.

Addition

Proton transfer

Migration
(key step!)

Proton transfer

Minor note: the rate of the addition reaction is increased by adding base, which deprotonates the peroxyacid, making it a better nucleophile.

m-chloro benzoic acid (byproduct)

Mg
Magnesium

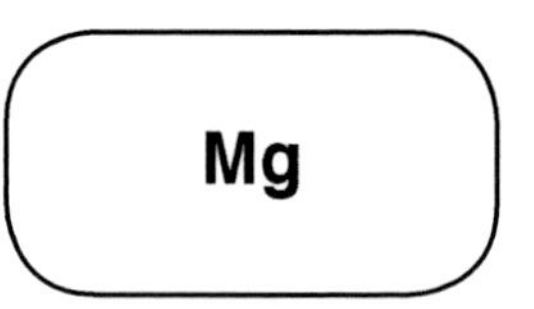

Also known as: Mg^0, Mg(s)

What it's used for: Magnesium metal is used for the formation of Grignard reagents from alkyl and alkenyl halides. A common solvent for these reactions are ethers, such as diethyl ether (Et_2O).

Similar to: Lithium (in the formation of alkyllithium reagents), Na, K

Example 1: Conversion of alkyl halides to Grignard reagents

Br — Mg / ether → MgBr

Diethyl ether is a common solvent for this reaction

Example 2: Conversion of alkenyl halides to Grignard reagents

Cl — Mg / ether → MgCl

How it works: ***Formation of Grignard reagents***

Although the mechanism for Grignard formation is generally not given in text-books, formation of Grignard reagents goes through a radical process.

Br ·Mg(s) → · MgBr → MgBr

Here, magnesium donates a single electron to Br, which forms a radical anion. Homolytic fragmentation of the C–Br bond leads to a free radical, which recombines with MgBr to give the Grignard reagent.

MsCl
Methanesulfonyl chloride

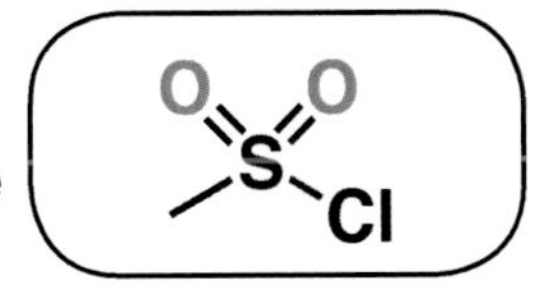

Also known as: Mesyl chloride

What it's used for: Methanesulfonyl chloride converts alcohols into good leaving groups. It behaves essentially identically to TsCl for this purpose.

Similar to: p-toluenesulfonyl chloride (TsCl)

Example 1: Conversion of alcohols to alkyl mesylates

OH → (MsCl) → OMs

pyridine (a base)

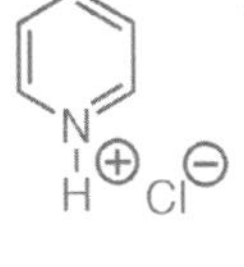

It is common to use a weak base (pyridine in this example) to react with the HCl that is formed as a byproduct of this reaction. This helps the reaction proceed to completion.

The resulting alkyl sulfonates ("mesylates") are excellent leaving groups in substitution and elimination reactions. For mechanisms and examples, see the section on TsCl.

NaN_3
Sodium azide

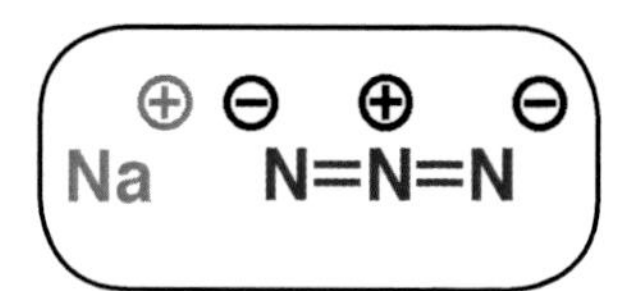

What it's used for: Sodium azide is a good nucleophile that readily participates in S_N2 reactions. Alternatively the sodium or lithium salt of azide ion can be used, but sodium is the most common.

Similar to: LiN_3, KN_3

Example 1: Substitution reaction - conversion of alkyl halides to alkyl azides

TsO → (NaN_3, DMSO) → N_3 + NaOTs

Dimethyl sulfoxide (DMSO) is a common solvent for S_N2 reactions

How it works: ***Nucleophilic substitution***

Sodium azide is the conjugate base of the weak acid HN_3 (pKa = 4.7). It is an excellent nucleophile that happens to be a weak base; reactions using N_3 will have very little competition from elimination pathways.

Na :N=N=N: + OTs → N=N=N: + NaOTs

Na
Sodium

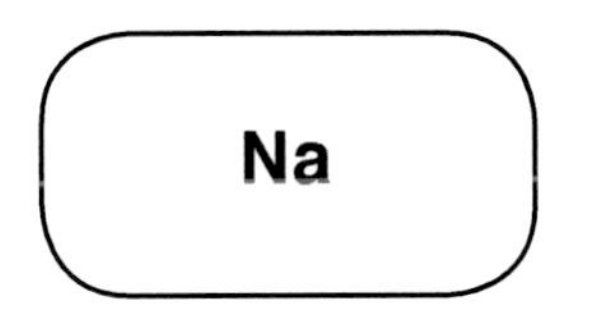
Na

What it's used for: Sodium is a very strong reducing agent. It will reduce alkynes to alkenes (trans-alkynes, specifically). It will form hydrogen gas when added to alcohols, resulting in alkoxides. It will also reduce aromatic groups to alkenes (the Birch reduction).

Similar to: Lithium (Li), potassium (K)

Example 1: Reduction - conversion of alkynes to *trans* alkenes

Na / NH_3

Example 2: Conversion of alcohols to alkoxides

OH —Na→ ONa + 1/2 H_2

Example 3: Birch reduction - conversion of arenes to dienes

Na / NH_3 / *t*-BuOH

How it works: ***Reduction of alkynes to trans-alkenes***

Sodium metal has an extremely low ionization energy and will readily give up its electron.

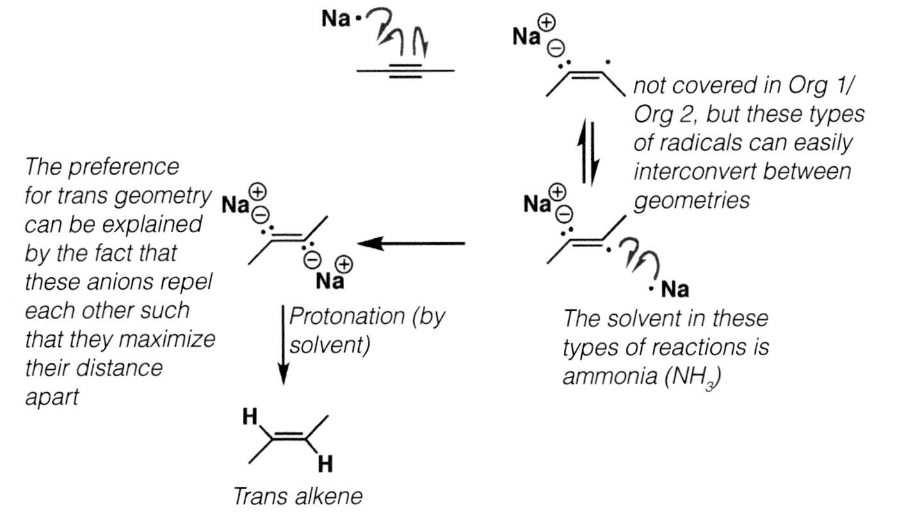
Na·
Na
not covered in Org 1/ Org 2, but these types of radicals can easily interconvert between geometries
The preference for trans geometry can be explained by the fact that these anions repel each other such that they maximize their distance apart
Na
Na
Na
·Na
Protonation (by solvent)
The solvent in these types of reactions is ammonia (NH3)
H
H
Trans alkene

Na
(continued)

How it works: ***The Birch Reduction***

The Birch reduction is a method for transforming aromatic rings into dienes. The products obtained depend on the substituent. Electron donating groups provide products where the alkene is attached to the electron donating group (enol ethers):

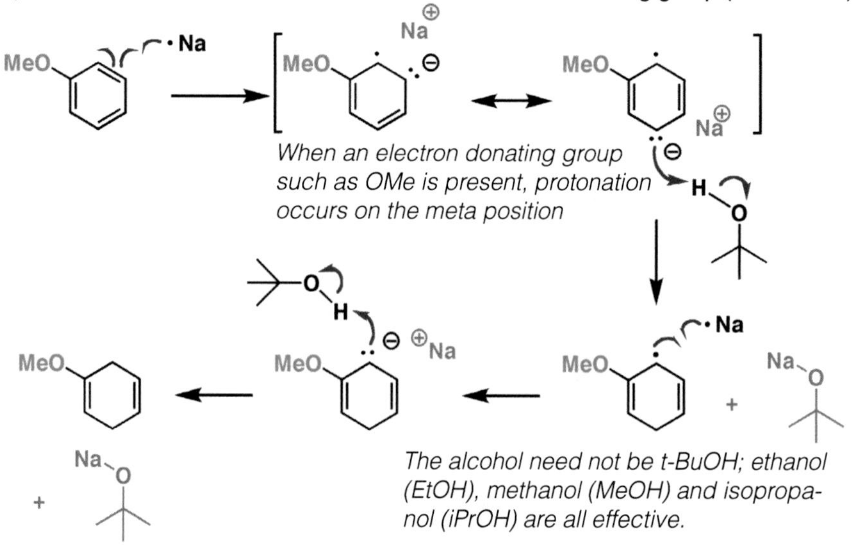

When an electron donating group such as OMe is present, protonation occurs on the meta position

The alcohol need not be t-BuOH; ethanol (EtOH), methanol (MeOH) and isopropanol (iPrOH) are all effective.

When electron withdrawing substituents are present, protonation occurs on the carbon adjacent to the electron withdrawing group

Note that protonation occurs adjacent to the electron withdrawing group

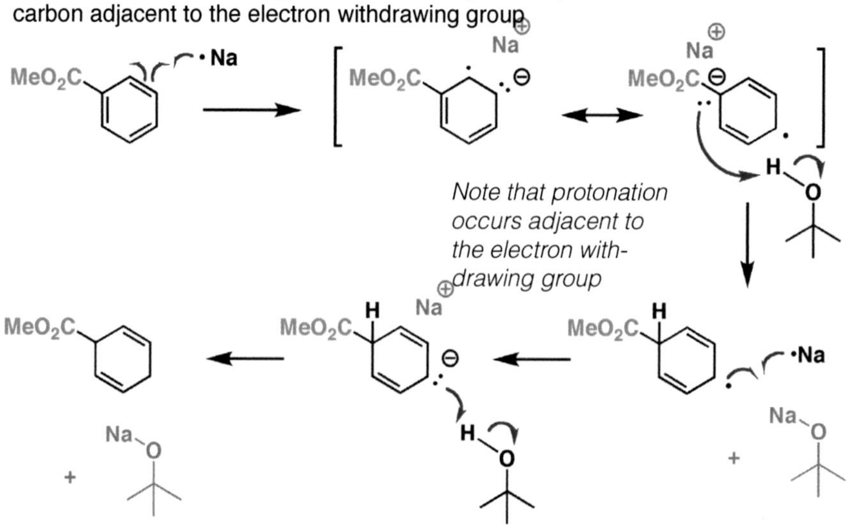

$NaBH_4$
Sodium borohydride

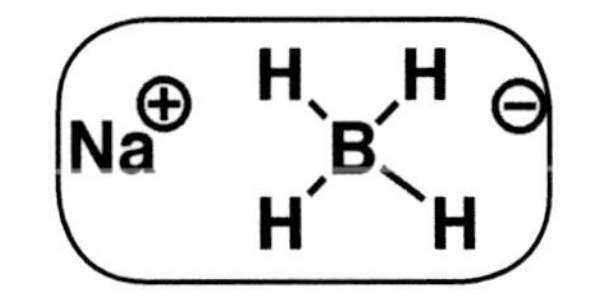

What it's used for: Sodium borohydride is a reagent mainly used for the reduction of ketones and aldehydes (it will also reduce acid halides). It is also used in the oxymercuration of alkenes, to replace mercury with H.

Similar to: Lithium aluminum hydride ($LiAlH_4$) but less reactive. Also similar to other reducing agents such as $NaCNBH_3$, DIBAL, $LiAlH(Ot\text{-}Bu)_3$, etc.

Example 1: Reduction - conversion of ketones to secondary alcohols

O → ($NaBH_4$) → OH

Example 2: Reduction of aldehydes to primary alcohols

O, H → ($NaBH_4$) → OH

Example 3: Oxymercuration - conversion of alkenes to alcohols

1) $Hg(OAc)_2$, H_2O
2) $NaBH_4$
→ OH

How it works: ***Reductions of aldehydes and ketones***

Sodium borohydride is a good reducing agent. Although not as powerful as $LiAlH_4$, it is very effective for the reduction of aldehydes and ketones to alcohols. It will generally not reduce esters or amides.

Na⊕ ⊖ BH_4 + O=CH(propyl) — *Addition* → Na⊕⊖:Ö: (alkoxide) + BH_3

Protonation ↓

OH (primary alcohol)

This reaction is commonly performed in a solvent such as CH_3OH, which serves as a source of protons.

$NaBH_4$ (continued)

How it works: ***In the oxymercuration reaction***

In the oxymercuration reaction, $NaBH_4$ is used to break the carbon-mercury bond and replace it with hydrogen (second step)

Hg(OAc)$_2$, H_2O → HgOAc / OH + HOAc

HgOAc / OH —$NaBH_4$→ OH + Hg(s) + BH_3 + NaOAc

The mechanism for this process is **rarely covered in textbooks** but is covered here for the sake of completeness.

Na⊕ ⊖ H–B(H)(H)–H → Hg–OAc / OH → H–Hg / OH + NaOAc + BH_3

Hydride attacks Hg, displacing the acetate ion

The C–Hg bond is weakened and breaks homolytically to give a carbon radical

H–Hg / OH

The carbon radical then abstracts hydrogen from the resulting Hg(I), giving elemental mercury as a product.

Hg (liquid) + H / OH

Net result is replacement of mercury for hydrogen.

The deuterated version of $NaBH_4$, sodium borodeuteride ($NaBD_4$) will replace Hg with D.

HgOAc / OH —$NaBD_4$→ D H / OH + Hg(s) + BD_3 + NaOAc

$NaBH(OAc)_3$
Sodium triacetoxy borohydride

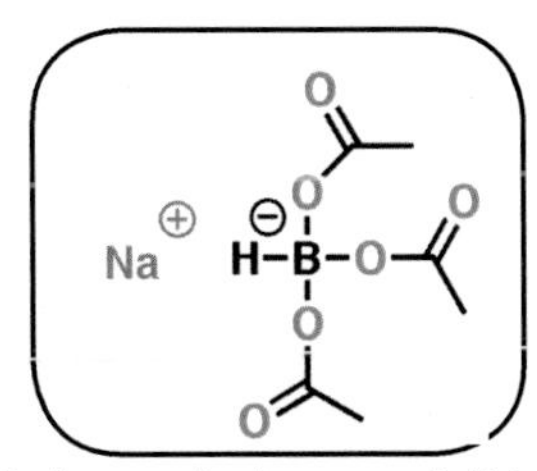

What it's used for: Sodium triacetoxy borohydride is a reducing agent. It is most often used for the reductive amination of ketones (and aldehydes) to amines. In this respect it is identical to sodium cyanoborohydride ($NaCNBH_3$). Reduction is often performed under mildly acidic conditions.

Similar to: $NaCNBH_3$, $NaBH_4$

Example 1: Reductive amination - conversion of aldehydes to amines

NH_2 / $NaBH(OAc)_3$ → HN

Example 2: Reductive amination - conversion of ketones to amines

NH_2 / $NaBH_3CN$ → HN

How it works: ***Reductive amination***

Reductive amination with $NaBH(OAc)_3$ works essentially the same way as it does for sodium cyanoborohydride ($NaBH_3CN$)

$NaCNBH_3$

Sodium cyanoborohydride

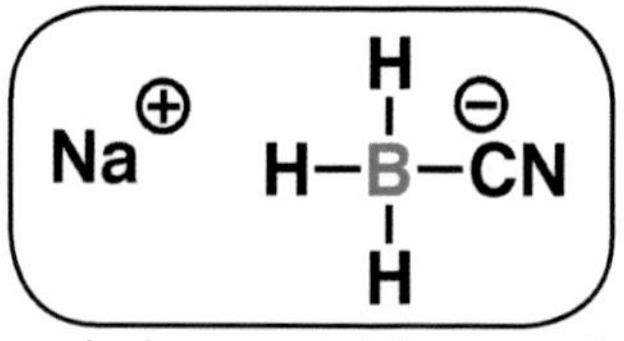

What it's used for: Sodium cyanoborohydride is a reducing agent. It is generally used for reductive amination - the reduction of imines (or "iminiums") to amines. It's common to perform this reaction under slightly acidic conditions (pH 4-5)

Similar to: $NaBH_4$, $NaBH(OAc)_3$

Example 1: Reductive amination - conversion of ketones to amines

NH_2 / $NaBH_3CN$ / pH 4.5 (mildly acidic) → HN

Example 2: Reductive amination - conversion of aldehydes to amines

NH_2 / $NaBH_3CN$ / pH 4.5 (mildly acidic) → HN

How it works: ***Reductive amination***

The first step is formation of an imine from the aldehyde/ketone and the amine:

NH_2 / $NaBH_3CN$ / pH 4-5 → *imine* + H_2O

At pH 4-5 a significant proportion of the imine is present as its conjugate acid ("iminium") which is a more reactive electrophile than the imine.

imine ⇌ (pH 4.5) *iminium* → *amine* + H–B(H)–CN

H–B(H)(H)–CN ⊖ Na ⊕

Interestingly, $NaCNBH_3$ is a poorer reducing agent than $NaBH_4$. It is used because at slightly acidic pH (~4-5) it is selective for reducing iminium ions (the conjugate acids of imines) over aldehydes and ketones.

The process of converting a ketone or aldehyde to an amine in the presence of a reducing agent such as $NaBH_3CN$ is called "reductive amination".

NaH
Sodium Hydride

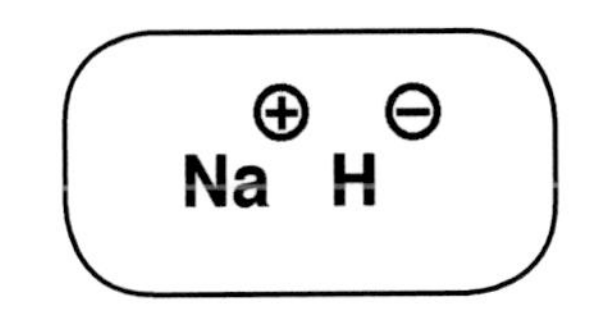

What it's used for: Sodium hydride is a strong base and a poor nucleophile. It is useful for deprotonating alcohols and alkynes, among others. One advantage of using NaH is that the byproduct is H_2, a gas which does not further interfere with the reaction.

Similar to: Potassium hydride (KH), lithium hydride (LiH)

Example 1: Acid-base reaction - conversion of alkynes to acetylides

H —NaH→ ⊖ ⊕Na

+ 1/2 H_2

Example 2: Acid-base reaction - conversion of alcohols to alcohols

OH —NaH→ ⊖O ⊕Na + 1/2 H_2

Example 3: Deprotonation of phosphonium salts to form ylides

H H R ⊕PPh$_3$ Br⊖ —NaH→ H ⊖ R ⊕PPh$_3$ + NaBr + 1/2 H_2

How it works: ***Deprotonation***

H^- is a strong base, the conjugate base of hydrogen (pKa = 42)

It will readily deprotonate alcohols (pKa 16-18), alkynes (pKa 25), and other species that are more acidic than hydrogen.

One advantage of using NaH (and KH) is that the conjugate acid is a gas (H_2) and will bubble out of the reaction vessel, not interfering with the reaction further. This also means that the deprotonation is irreversible.

For most purposes it can be used interchangably with $NaNH_2$ and other strong bases.

$NaIO_4$
Sodium periodate

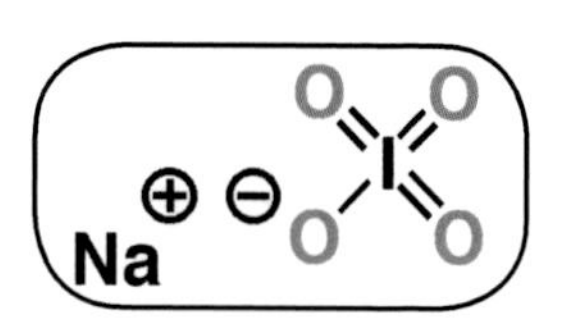

What it's used for: Sodium periodate is a strong oxidant. It will cleave 1,2 diols (vicinal diols) to give aldehydes and ketones.

Similar to: Periodic acid (HIO_4), Lead (IV) acetate [$Pb(OAc)_4$]

Example 1: Oxidative cleavage - conversion of diols to aldehydes/ketones

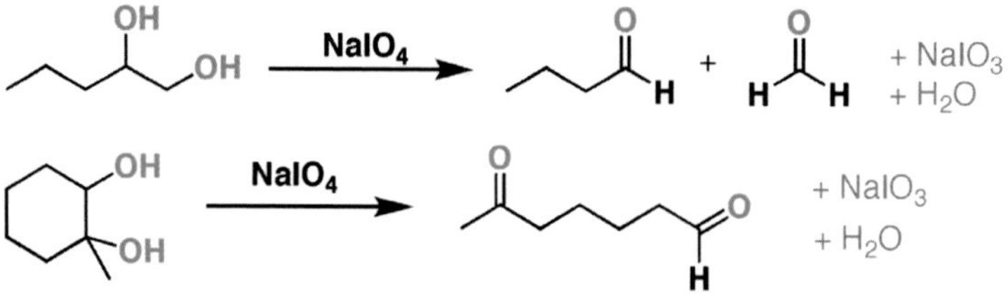

OH, OH —$NaIO_4$→ O, H + O, H H + $NaIO_3$ + H_2O

OH, OH —$NaIO_4$→ O, O, H + $NaIO_3$ + H_2O

How it works: ***Oxidative cleavage of diols to give aldehydes/ketones***

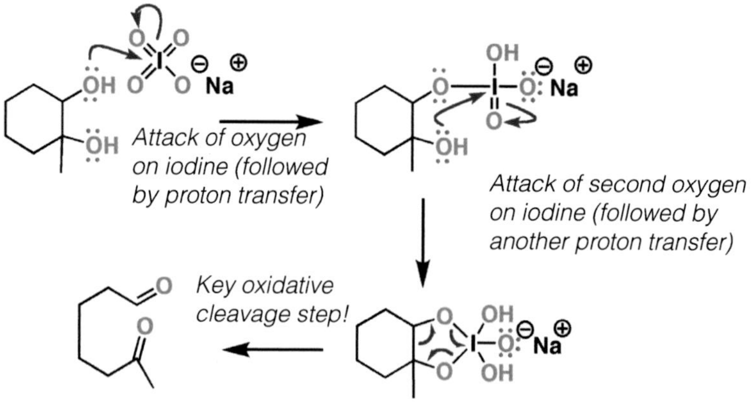

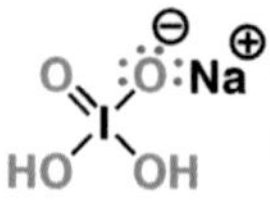

Attack of oxygen on iodine (followed by proton transfer)

Attack of second oxygen on iodine (followed by another proton transfer)

Key oxidative cleavage step!

Trivial detail: this usually loses water to give $NaIO_3$ plus H_2O

Note that in this process iodine (VII) has been reduced to iodine (V)

$NaNH_2$
Sodium amide

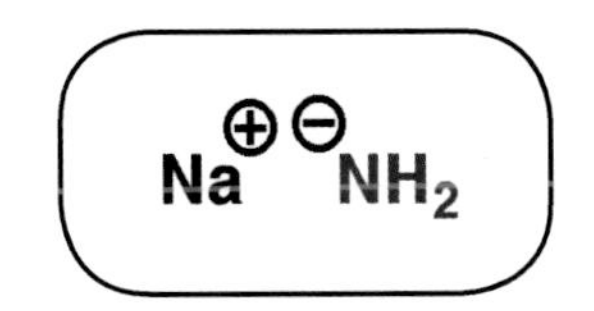

Also known as: Sodamide

What it's used for: Sodium amide is a very strong base, useful for the deprotonation of alkynes and also in elimination reactions toward the formation of alkynes from dihalides. It can also be used to generate arynes ("benzynes") which can undergo nucleophilic attack.

Similar to: $LiNH_2$, KNH_2. Essentially the same base strength as LDA, although less sterically hindered.

Example 1: Acid-base reaction - conversion of alkynes to acetylides

H —($NaNH_2$)→ ⊖ ⊕ Na

$+NH_3$

Example 2 - Elimination - conversion of geminal dihalides to alkynes

Br Br —($NaNH_2$ (2 equiv))→ + 2 NaBr

"Geminal" dihalide - has two halogens on the same carbon

Example 3: Acid-base reaction - conversion of vicinal dihalides to alkynes

Br Br —($NaNH_2$ (2 equiv))→ + 2 NaBr

"Vicinal" dihalide - has two halogens on adjacent carbons

Example 4: Conversion of aryl halides to aryl amines (via arynes)

H Br —($NaNH_2$ / NH_3)→ H NH_2 + NaBr

This proceeds through an aryne mechanism

How it works: ***As a strong base***

$NaNH_2$ is the conjugate base of ammonia (pKa 38). It is sufficiently strong to deprotonate alkynes, which cannot be done reliably with NaOH.

$NaNH_2$ is also a useful reagent for performing the elimination of geminal dihalides to alkynes.

Br Br H H H Na⊕ :NH_2⊖ —(*E2 reaction*)→ Br H H :NH2⊖ ——→ H

$+ NH_3 + NaBr$ $+ NH_3 + NaBr$

NaOH

Sodium hydroxide

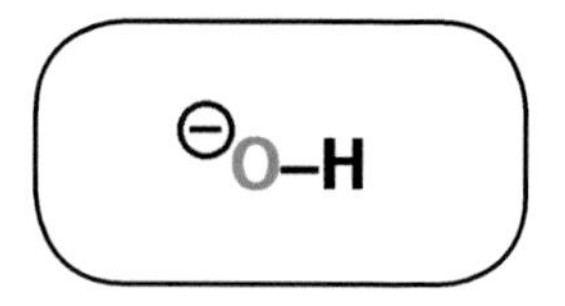

What it's used for: Hydroxide ion (often encountered as NaOH or KOH) is a strong base and good nucleophile. It is impossible to mention all of its applications here but a few of its crucial reactions are highlighted.

Similar to: Similar in action to other strong bases.

Example 1: Acid-base reaction - conversion of alcohols to alkoxides

R O H — Na OH → R O Na + H_2O

Example 2: Elimination - conversion of alkyl halides to alkenes

Br — Na OH, H_2O, Δ (heat) → + NaBr + H_2O

Example 3: Acid-base reaction - conversion of ketones/aldehydes to enolates

O H H H — Na OH → O Na H H + H_2O

Example 4: Substitution - conversion of alkyl halides to alcohols

Br — Na OH → OH + KBr

This is an S_N2 reaction

Example 5: Acyl substitution - conversion of acyl halides to carboxylic acids

O Cl — Na OH → O OH + NaBr

Example 6: Acyl substitution (saponification) - conversion of esters to carboxylic acids

O OCH_3 — Na OH → O OH + $NaOCH_3$

Example 7: Acyl substitution - conversion of anhydrides to carboxylic acids

O O O — Na OH → O OH + O O

NBS
N–Bromosuccinimide

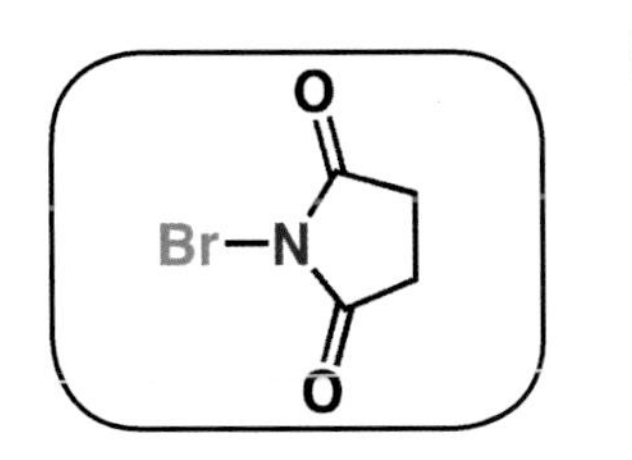

What it's used for: NBS is a source of reactive, electrophilic bromine. It is often used for allylic bromination and in formation of halohydrins from alkenes. Since it is a crystalline solid it is more convenient to use than liquid elemental Br_2.

Similar to: Br_2, NCS (*N*-chlorosuccinimide), NIS (*N*-iodosuccinimide)

Example 1: Allylic bromination - conversion of alkenes to allylic bromides

NBS / hν (light) → Br + HBr + H–N (succinimide)

This byproduct is called "succinimide"

succinimide

Example 2: Conversion of alkenes to bromohydrins

NBS / H_2O → Br, OH + H–N (succinimide)

Bromohydrin

succinimide

How it works: ***Halohydrin formation***

As with Br_2, alkenes treated with NBS will form bromonium ions:

Br—N → [Br: ⊕, ⊖ :N] —H_2O→ Br, OH

Bromonium ion formation

Trans product

The trans product is formed exclusively due to attack of nucleophile on the face opposite the bromonium ion. Note - in this case a 1:1 mixture of enantiomers is formed.

Br: ≡ Br: → Br, HO⊕ H, :N⊖ → Br, OH

Side view :ÖH₂

Attack by solvent (water in this case)

Deprotonation

Top view

+ succinimide

Br, OH

(Enantiomer)

NBS
(continued)

How it works: ***Allylic bromination***

NBS provides a constant, low concentration of Br_2, which is formed when HBr (from propagation step 1) reacts with NBS. This is useful because the low concentration of Br_2 prevents dibromination of the double bond from occuring.

Generation of Br_2

O O N–Br —HBr→ N–H + Br–Br O O

Initiation step

Br–Br ⟶ 2 Br•

When bromine is heated or treated with light, homolytic cleavage of the Br–Br bond results in the formation of two bromine radicals.

Propagation step 1

In the first propagation step, the bromine radical removes a hydrogen from the allylic carbon, giving a resonance-stabilized free radical.

Br• ⟶ Br• H ⟶ + HBr

Propagation step 2

In the second propagation step, the allylic radical reacts with Br_2, giving the allylic bromide and regenerating a new Br radical. This continues the catalytic cycle.

—Br–Br→ Br–Br ⟶ Br + Br•

NCS
N–Chloro Succinimide

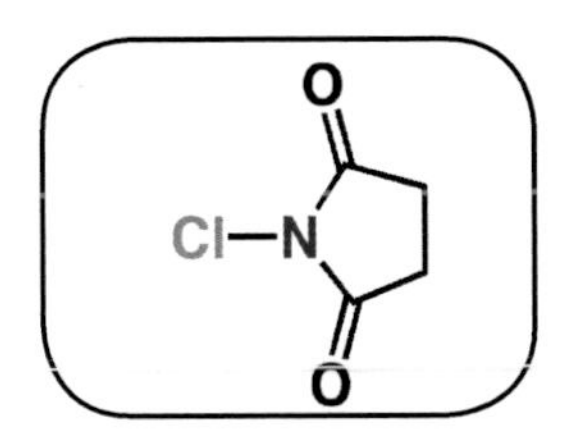

What it's used for: N-chlorosuccinimide is a source of reactive (electrophilic) chlorine. It is used for the formation of chlorohydrins from alkenes. A crystalline solid, it is more easily handled than dangerous chlorine gas.

Similar to: Cl_2, *N*-bromosuccinimide (NBS), *N*-iodosuccinimide (NIS)

Example 1: Conversion of alkenes to chlorohydrins

NCS, H_2O → Cl, OH + H–N (succinimide)

Chlorohydrin *succinimide*

How it works: ***Formation of chlorohydrins***

The first step in chlorohydrin formation is attack of the nucleophilic alkene upon the electrophilic chlorine. This forms an intermediate chloronium ion. In the presence of a nucleophilic solvent such as H_2O, this chloronium ion is attacked to give the chlorohydrin.

Cl–N → [Cl: , :N] → H_2O → Cl, OH

Chloronium ion

Since attack on the chloronium ion occurs exclusively from the back side, the trans product is formed exclusively.

Cl: ≡ Cl: :OH_2 → Cl, HO, H, :N → Cl, OH

Note - the enantiomer is also formed here

NIS

N–Iodo Succinimide

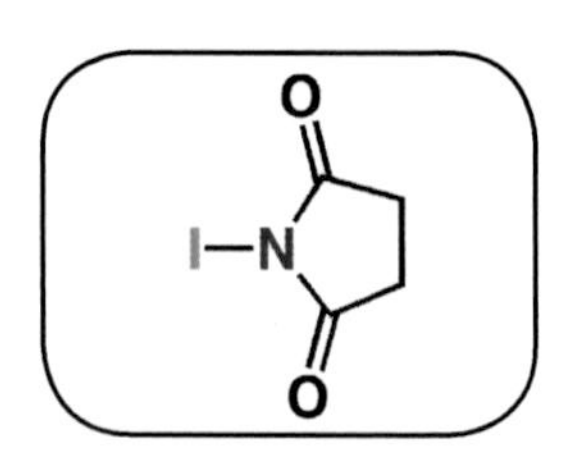

What it's used for: N-iodosuccinimide is a source of electrophilic iodine, similar to NBS and NCS. When added to an alkene in the presence of water, it will form iodohydrins.

Similar to: I_2, NBS, NCS.

Example 1: Conversion of alkenes to iodohydrins

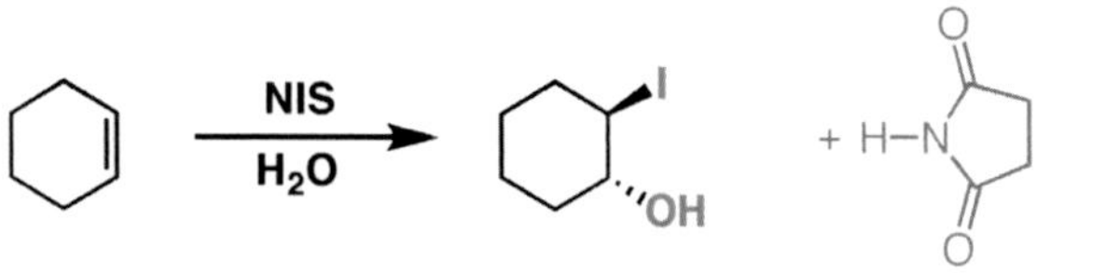

NIS, H_2O → Iodohydrin + succinimide

How it works: ***Iodohydrin formation***
The reaction proceeds via attack of the alkene upon the iodine, followed by attack of nucleophilic solvent upon the iodonium ion.

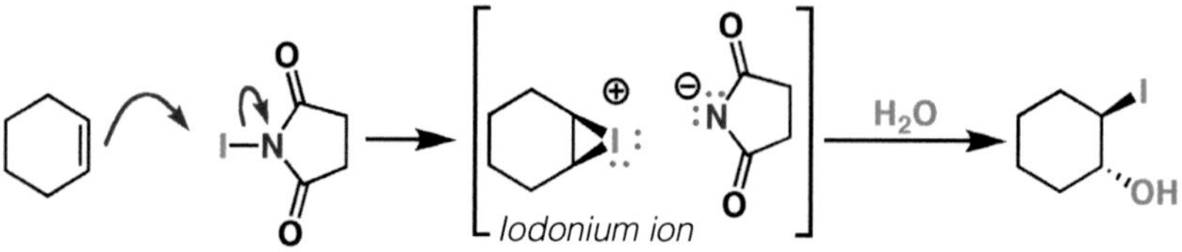

Iodonium ion

Since attack on the iodonium ion occurs exclusively from the back side, the trans product is formed exclusively.

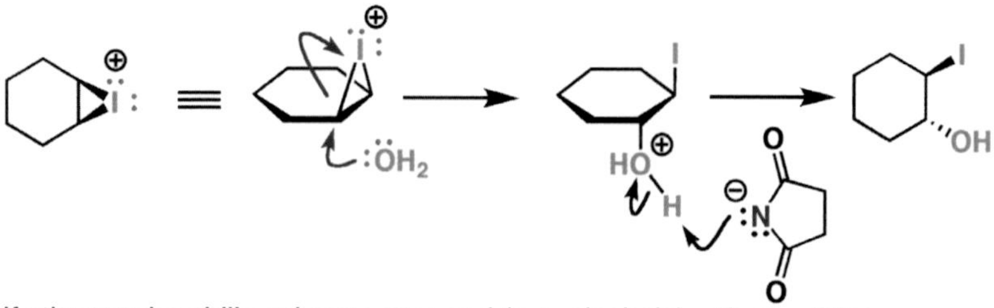

If other nucleophilic solvents are used (e.g. alcohols), ethers will be formed. This is an example of *iodoetherification.*

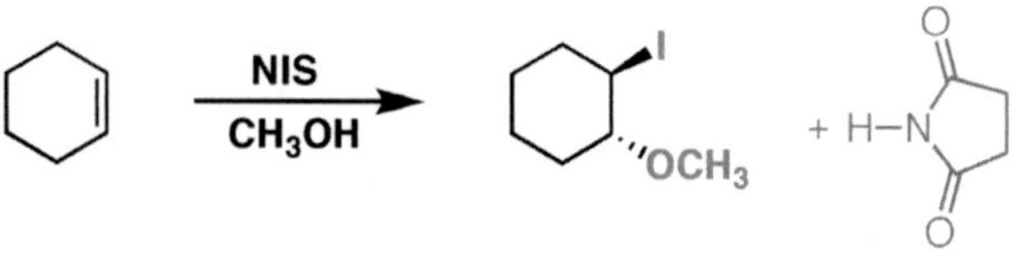

NIS, CH_3OH → OCH_3 product + H–N succinimide

NH_2OH
Hydroxylamine

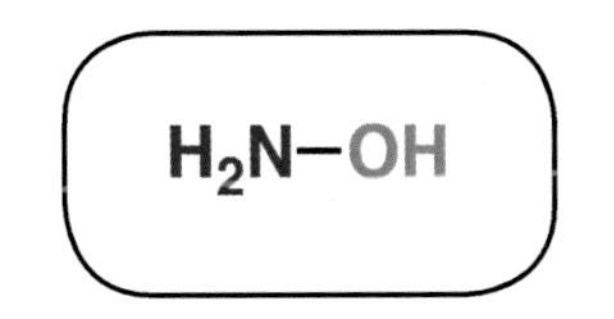

What it's used for: Hydroxylamine is a good nucleophile. It is most commonly used for the formation of oximes, a precursor to the Beckmann rearrangement.

Example 1: Conversion of ketones/aldehydes to oximes

NH2OH → Oxime + H2O

Example 2: Beckmann rearrangement - conversion of oximes to amides

NH2OH; H3O⊕, Δ (heat)

How it works: ***Conversion of ketones/aldeydes to oximes***

Treatment of an aldehyde or ketone with NH_2OH leads to formation of an oxime. Mild acid (although not shown) can accelerate this reaction.

:NH2OH — Addition → Proton transfer → Elimination → Deprotonation → Oxime + H2O

How it works: ***Beckmann rearrangement***

Treatment of the oxime with acid and heat leads to a rearrangement occurring simultaneously with loss of water. The product is a nitrile.

Oxime — Protonation → Rearrangement → Deprotonation → R—C≡N + HCl

Nitrile

NH_3
Ammonia

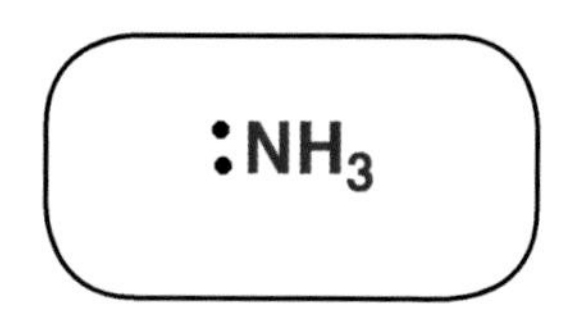

Also known as: NH_3 (l) (signifies it is a liquid)

What it's used for: Ammonia is a base and a nucleophile. It is often used as a solvent in reactions involving lithium (Li), sodium (Na), and potassium (K). It has quite a low boiling point (–33° C)

Example 1: As a solvent - conversion of alkynes to acetylides

Cyclohexyl–CH₂–C≡C–H $\xrightarrow[NH_3]{NaNH_2}$ Cyclohexyl–CH₂–C≡C:⁻ ⁺Na

Example 2: As a nucleophile - conversion of acyl chlorides to amides

$CH_3COCl \xrightarrow{NH_3} CH_3CONH_2$ + HCl

How it works:

NH_3 is the simplest amine and is a Lewis base due to its unshared lone pair of electrons.

Being the conjugate acid of $NaNH_2$, it is the perfect solvent for this base, much like MeOH is used as a solvent for NaOMe.

NH_2NH_2
Hydrazine

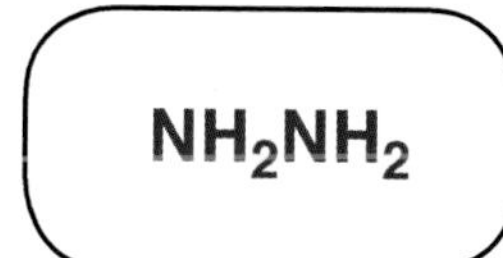

What it's used for: Hydrazine is a good reductant and nucleophile. It is used in the Wolff-Kishner reaction, a means of converting ketones to alkanes. It is also used in the final step of the Gabriel amine synthesis to liberate the free amine.

Example 1: Wolff-Kishner reaction - conversion of ketones to alkanes

NH_2NH_2, NaOH, HO–CH₂CH₂–OH, Δ (heat) → cyclohexane $+ N_2 + H_2O$

ethylene glycol (a high boiling solvent)

Example 2: Gabriel synthesis - conversion of phthalimide to primary amine

N-propylphthalimide —NH_2NH_2→ H_2N–propyl

How it works: ***The Wolff-Kishner reaction***

Hydrazine, like other amines, will add to aldehydes and ketones to form condensation products. The imine of hydrazine is called a "hydrazone".

cyclohexanone —NH_2NH_2, HO–CH₂CH₂–OH→ *hydrazone* $+ H_2O$

When treated with strong base and heated vigorously, nitrogen gas is liberated and an alkane is formed. This is the Wolff-Kishner reaction

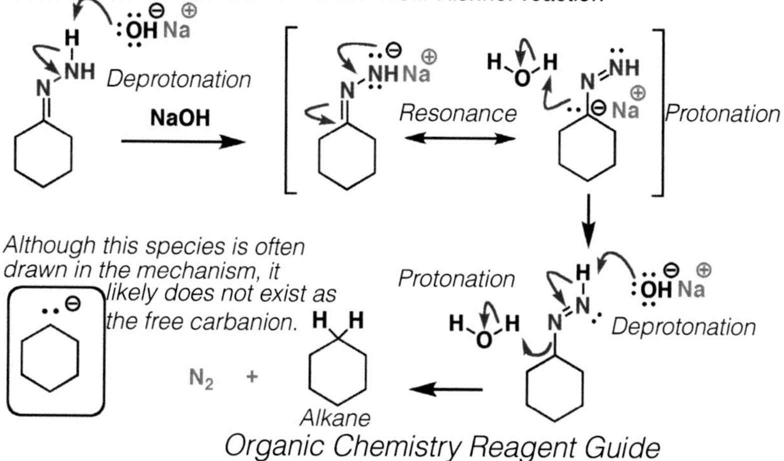

Ni_2B
Nickel boride

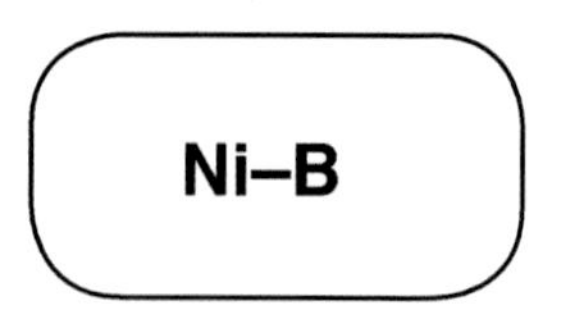

What it's used for: Nickel boride is a reducing agent, used for the reduction of alkynes to give *cis*-alkenes.

Similar to: Lindlar's catalyst

Example 1: Reduction - conversion of alkynes to alkenes

R—≡—R —(Ni–B, H_2)→ cis-RHC=CHR

How it works: ***Partial reduction of alkynes.***

Nickel boride is generally formed in the reaction vessel by adding $NaBH_4$ to nickel (II) salts such as $NiCl_2$. It behaves the same as Lindlar's catalyst, performing partial hydrogenation of alkynes to give alkenes. The stereochemistry is always *syn.*

It can also be used in the presence of hydrogen gas as in the example above.

Nickel boride does not react with alkenes.

O_3
Ozone

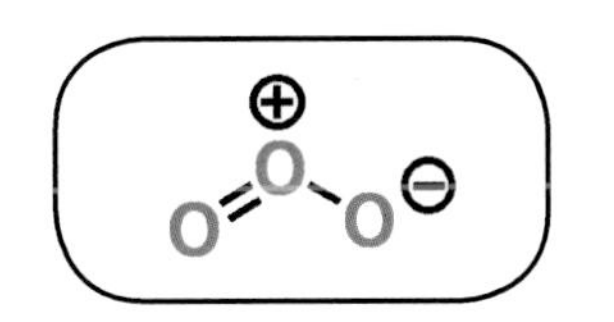

What it's used for: Ozone is an oxidizing agent. It will cleave alkenes and alkynes to give carbonyl compounds, in a reaction called "oxidative cleavage". The products formed can be dependent on the type of workup used. Reductive workup preserves aldehydes, whereas oxidative workup will oxidize any aldehydes to carboxylic acids.

Similar to: $KMnO_4$

Example 1: Oxidative cleavage (reductive workup) - conversion of alkenes to aldehydes/ketones

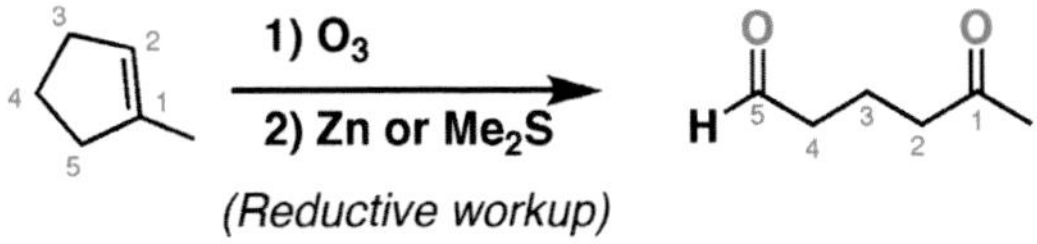

1) O_3
2) Zn or Me_2S
(Reductive workup)

Example 2: Oxidative cleavage (oxidative workup) - conversion of alkenes to carboxylic acids/ketones

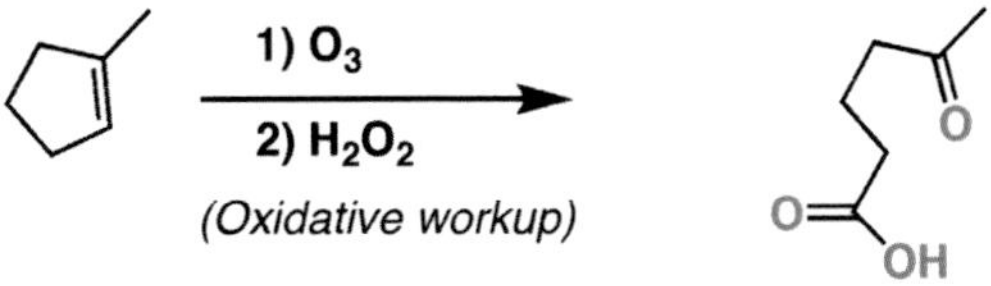

1) O_3
2) H_2O_2
(Oxidative workup)

Example 3: Oxidative cleavage - conversion of alkynes to carboxylic acids

1) O_3
2) H_2O_2

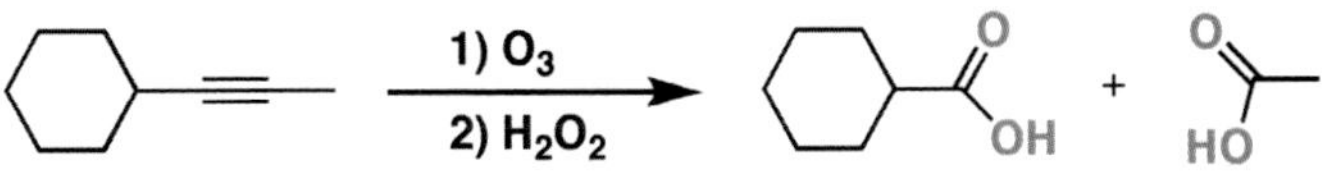

How it works: ***Oxidative cleavage of alkenes***

Treatment of an alkene with ozone leads to initial formation of a molozonide, followed by rearrangement to an ozonide. Reduction gives the carbonyl products.

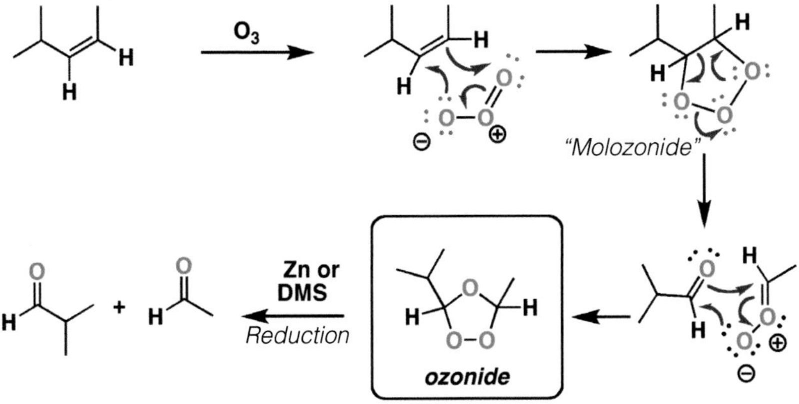

"Molozonide"

Zn or DMS
Reduction

ozonide

O_3 (continued)

How it works: ***Reductive workup***

Dimethyl sulfide is a reducing agent. It breaks the weak O–O bond, and expels dimethyl sulfoxide (DMSO)

DMSO

How it works: ***Oxidative workup***

Hydrogen peroxide oxidizes aldehydes to carboxylic acids. Upon warming, the ozonide opens up to an aldehyde and carbonyl oxide, which are converted to carboxylic acids.

Ozonide (decomposes upon warming)

H_2O_2

H_2O

Addition

Proton transfer

Deprotonation

Trivial detail - the deprotonation step is rate-limiting, so the reaction can be accelerated by the addition of a base such as NaOH

R_2CuLi
Organocuprates

R_2CuLi

Also known as: Gilman reagents

What it's used for: Organocuprate reagents (Gilman reagents) are carbon nucleophiles. They will perform [1,4] additions ("conjugate additions") to α,β unsaturated ketones, as well as S_N2 reactions with certain types of alkyl halides. They can also add to acyl halides to give ketones.

Similar to: Grignard reagents, organolithium reagents

Example 1: Conversion of alkyllithiums to organocuprates

Li (2 equiv) → CuI → ( )₂Cu⊖ Li⊕ + Li⊕ I⊖

Can also use CuCl, CuBr, CuCN

Example 2: Conjugate addition - addition to α, β unsaturated ketones

1) ( )₂Cu⊖ Li⊕
2) $H_3O^{\oplus}$

This reaction works best for methyl, allyl, and benzyl halides

Example 3: Substitution - conversion of alkyl halides to alkanes (S_N2)

CH_3–I → ( )₂Cu⊖ Li⊕ → CH_3 + Cu + LiI

Br → ( )₂Cu⊖ Li⊕ → + Cu + LiBr

Example 4: Acyl substitution - conversion of acyl halides to ketones

O, Cl → ( )₂Cu⊖ Li⊕ → O + LiCl + Cu

How it works: ***As a nucleophile***

δ− Cu⊖ δ+ Li⊕ δ−

Electronegativity	
C	2.5
Cu	1.9

Due to carbon's higher electronegativity relative to Cu, it bears a partial positive charge and is thus nucleophilic.

RLi
Organolithium reagents

R–Li

What it's used for: Organolithium reagents are extremely strong bases and good nucleophiles. they react with carbonyl compounds (aldehydes, ketones, esters, etc.) and epoxides. Being strong bases, they will also react with groups containing acidic hydrogens.

Similar to: Grignard reagents

Example 1: Conversion of alkyl halides to organolithiums

Br — Li (2 equiv) → Li + LiBr

Example 2: Conversion of aldehydes to secondary alcohols

Li — 1) O, H; 2) $H_3O^{\oplus}X^{\ominus}$ → OH + LiX

Acid is added in the second step to protonate the negatively charged oxygen

Example 3: Conversion of ketones to tertiary alcohols

Li — 1) O; 2) $H_3O^{\oplus}X^{\ominus}$ → OH + LiX

Acid is added in the second step to protonate the negatively charged oxygen

Example 4: Conversion of esters to tertiary alcohols

Li — 1) O, O; 2) $H_3O^{\oplus}X^{\ominus}$ → OH + HO + 2 LiX

Organolithium reagents add twice to esters, acid halides, and anhydrides

Example 5: Conversion of acyl halides to tertiary alcohols

Li — 1) O, Cl; 2) $H_3O^{\oplus}X^{\ominus}$ → OH + 2 RLi

Organolithium reagents add twice to esters, acid halides, and anhydrides

Example 6: Epoxide opening - conversion of epoxides to alcohols

Li — 1) O; 2) $H_3O^{\oplus}X^{\ominus}$ → OH + LiX

Organolithium reagents add to the less substituted end of epoxides. Acid is added to protonate the negatively charged oxygen.

R–Li
(continued)

Example 7: Reaction with carbon dioxide

1) (epoxide) 2) $H_3O^{\oplus}$ $X^{\ominus}$ + LiX

Example 8: Formation of organocuprates

(2 equiv) CuI $Cu^{\ominus}$ $Li^{\oplus}$ + LiI

Can also use CuCl, CuBr, or CuCN

Example 9: As a base

Organolithium reagents are extremely strong bases.

H H $\oplus$ $Br^{\ominus}$ PPh_3 Li PPh_3 + LiBr

ylide

Example 10: Addition to carboxylic acids

OH Li *(2 equiv)* + LiOH +

Organolithium reagents can add to carboxylic acids if 2 equivalents are added.

How it works: ***Addition to aldehydes/ketones***

Organolithium reagents are extremely strong nucleophiles. The electrons in the C–Li bond are highly polarized toward carbon:

δ^- Li δ^+ *behaves much like* $\ominus$ $\oplus$ Li

Organolithium reagents readily add to the electrophilic carbonyl atom in aldehydes and ketones. Subsequent addition of acid gives the neutral alcohol.

Addition $H^{\oplus}$ $X^{\ominus}$ $Li^{\oplus}$ *Protonation* OH + LiX

R–Li
(continued)

How it works: ***Addition to esters / acid halides / anhydrides***

Organolithium reagents add twice two these groups. The reaction proceeds through addition, elimination, and a second addition. Addition of acid at the end provies a neutral tertiary alcohol.

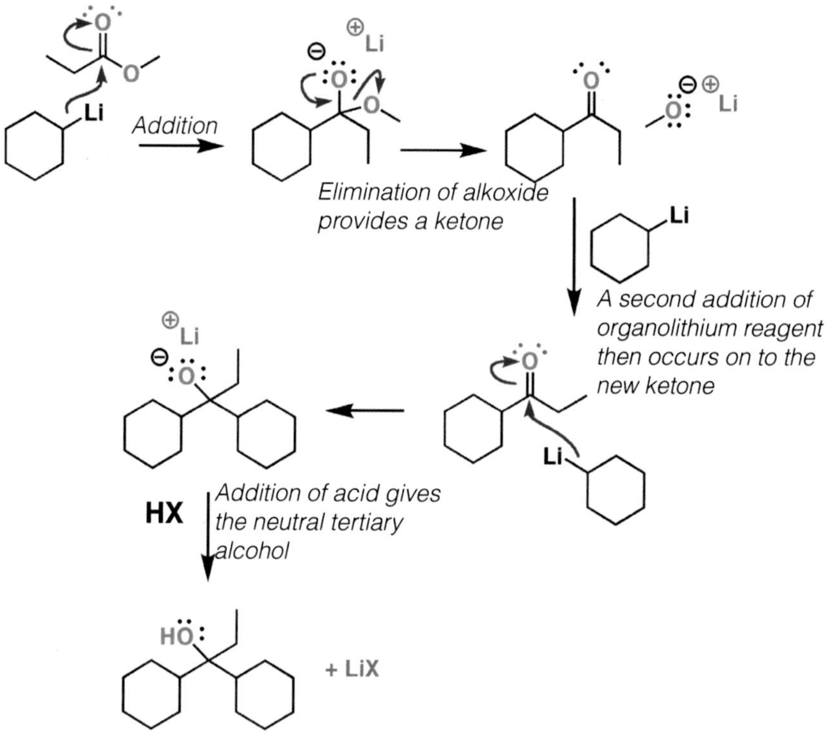

Addition

Elimination of alkoxide provides a ketone

A second addition of organolithium reagent then occurs on to the new ketone

HX *Addition of acid gives the neutral tertiary alcohol*

+ LiX

How it works: ***Addition to epoxides***

Organolithium reagents add to the least hindered end of epoxides (you can think of this like an S_N2). Protonation then gives the neutral alcohol.

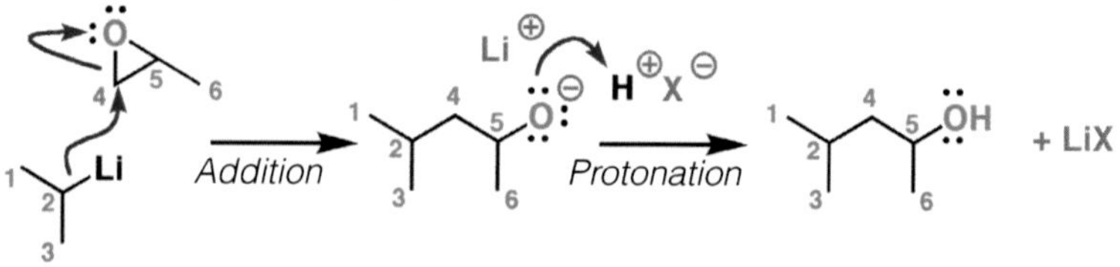

Addition *Protonation* + LiX

OsO_4
Osmium tetroxide

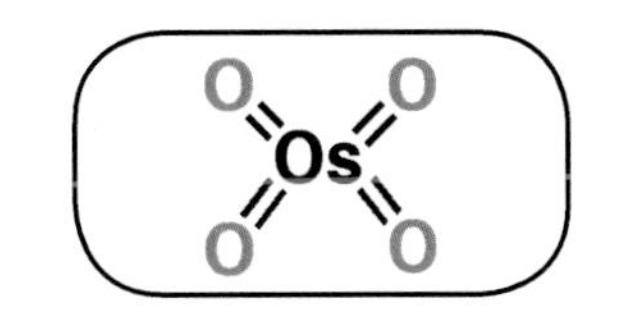

What it's used for: Osmium tetroxide is a reagent for the formation of 1,2-diols (vicinal diols) from alkenes. The selectivity for this reaction is always *syn.*

Similar to: $KMnO_4$ (cold, dilute)

Example 1: Dihydroxylation - conversion of alkenes to give vicinal diols

OsO_4

OH

OH

a "vicinal diol"

The stereochemistry of this reaction is always "syn"

OsO_4

$KHSO_3$

OH

OH

In the lab, $KHSO_3$ helps to remove the osmium from this reaction. Its presence has no effect on the final product (as shown here)

How it works: ***Dihydroxylation of alkenes***

Oxygens both approach the alkene from the same face.

O O Os O O

O O Os O O

$NaHSO_3$

OH

OH

Sometimes $NaHSO_3$ or $KHSO_3$ (bisulfite) is added to break down the cyclic osmium compound into a diol and an osmium salt.

$Pb(OAc)_4$
Lead tetraacetate

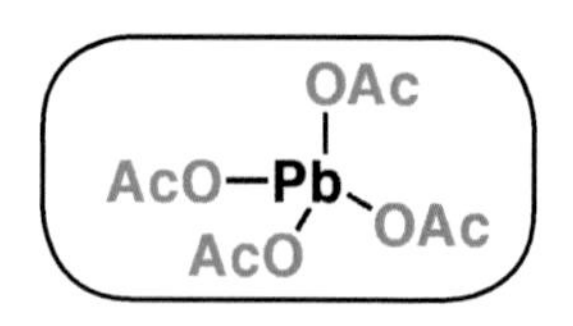

Also known as: Lead (IV) acetate

What it's used for: Lead tetraacetate will cleave 1,2-diols (vicinal diols) into aldehydes/ketones, similar to $NaIO_4$ and HIO_4.

Similar to: Sodium periodate ($NaIO_4$), periodic acid (HIO_4).

Example 1: Oxidative cleavage - conversion of diols to aldehydes / ketones.

$Pb(OAc)_4$ + $Pb(OAc)_2$ + 2 HOAc

$Pb(OAc)_4$ + $Pb(OAc)_2$ + 2 HOAc

How it works: ***Cleavage of diols to aldehydes/ketones***

Lead (IV) acetate is an oxidizing agent. It works by coordinating to the 1,2-diol and then breaking the carbon-carbon bond in a cyclic mechanism:

Coordination *(formed after proton transfer)* + HOAc

Coordination (and subsequent proton transfer) + HOAc

Oxidative cleavage + $Pb(OAc)_2$

Note that Pb(IV) has been reduced to Pb(II). The acetate groups are liberated as acetic acid.

PBr_3
Phosphorus tribromide

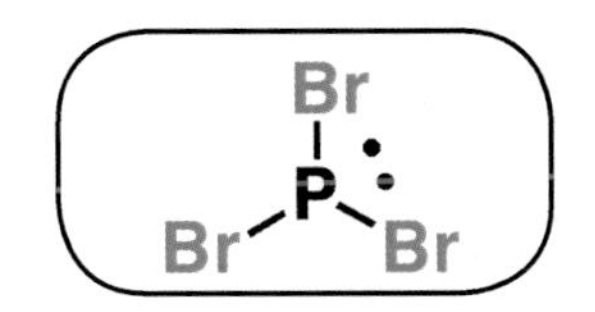

What it's used for: Phosphorus tribromide is a reagent for converting alcohols to alkyl bromides. It will also convert carboxylic acids to acid bromides (acyl bromides)

Similar to: Thionyl bromide ($SOBr_2$)

Example 1: Substitution - conversion of alcohols to alkyl bromides

PBr_3 OH → Br + $HOPBr_2$

Note: inversion

Example 2: Acyl substitution - conversion of carboxylic acids to acyl bromides

O OH → PBr_3 → O Br + $HOPBr_2$

How it works: ***Formation of alkyl bromides from alcohols***

Phosphorus tribromide is useful for converting alcohols into alkyl bromides with inversion of configuration. Driving force is formation of the strong P=O bond

How it works: ***Formation of acyl bromides***

Attack of oxygen on phosphorus

Addition

Elimination

Deprotonation

HBr + O=P–Br

PCl_3
Phosphorus trichloride

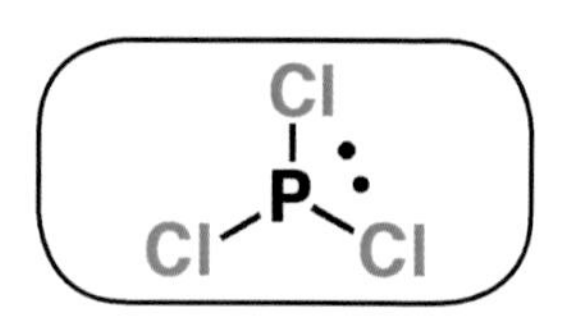

What it's used for: Phosphorus trichloride is a reagent for the conversion of alcohols to alkyl chlorides. It will also convert carboxylic acids to acid chlorides (acyl chlorides)

Similar to: $SOCl_2$, PCl_5. Mechanisms exactly the same as for PBr_3.

Example 1: Substitution - conversion of alcohols to alkyl chlorides

PCl_3 OH → Cl + $HOPCl_2$

Note: inversion

Example 2: Acyl substitution - conversion of carboxylic acids to acyl chlorides

O OH → PCl_3 → O Cl + $HOPCl_2$

How it works: *Formation of alkyl chlorides from alcohols*

OH Cl P Cl Cl → Cl P Cl O H Cl → Cl + Cl P O + HCl

How it works: *Formation of acyl chlorides from carboxylic acids*

Attack of oxygen on phosphorus → *Addition* → *Elimination* → *Deprotonation*

HCl + O P Cl + Cl O (benzoyl chloride)

PCl_5
Phosphorus Pentachloride

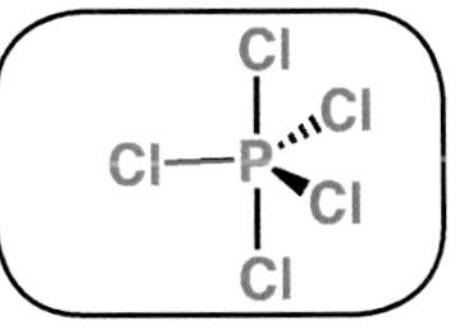

What it's used for: Phosphorus pentachloride will convert alcohols to alkyl chlorides, and carboxylic acids to acid chlorides (acyl chlorides)

Similar to: $SOCl_2$, PCl_3

Example 1: Formation of alkyl bromides from alcohols

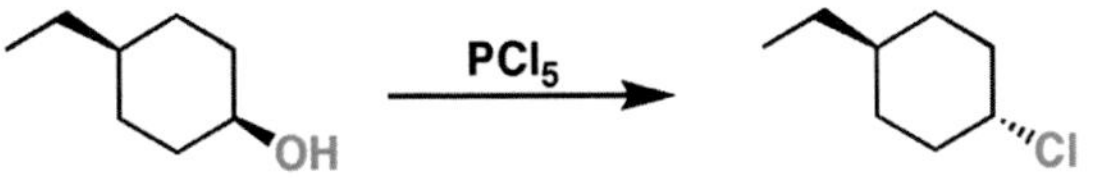

Note: inversion

Example 1: Formation of alkyl bromides from alcohols

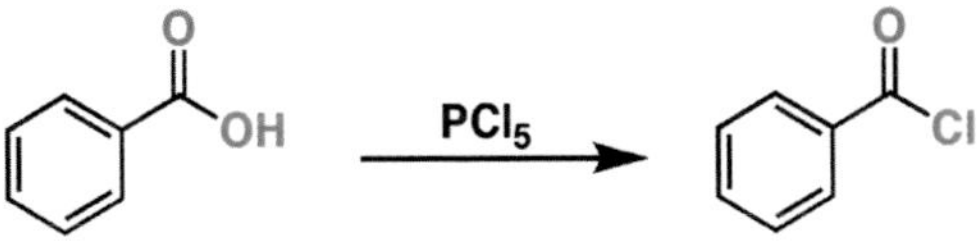

How it works: ***Formation of alkyl chlorides and acyl chlorides***

PCl_5 operates by a mechanism essentially identical to that of PCl_3 and PBr_3.

P_2O_5
Phosphus pentoxide

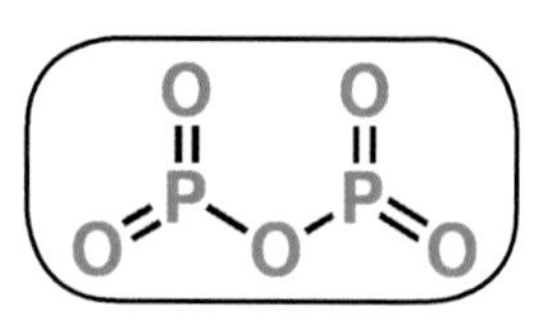

Also known as: Phosphoric anhydride; phosphorus (V) oxide; P_4H_{10}

What it's used for: P_2O_5 is a reagent used for dehydration. It is used for conversion of carboxylic acids to anhydrides, and also the formation of nitriles from amides.

Similar to: P_4O_{10} (this behaves exactly the same as P_2O_5)

Example 1: Conversion of carboxylic acids to anhydrides

P_2O_5

(2 equiv)

Example 2: Conversion of amides to nitriles

P_2O_5

How it works: ***Conversion of carboxylic acids to anhydrides***

Carboxylic acid #1

Proton transfer

Carboxylic acid #2

Elimination

Deprotonation

Anhydride

How it works: ***Conversion of amides to nitriles***

Amide

Addition

Proton transfer

Elimination

Deprotonation

Nitrile

http://masterorganicchemistry.com

Pd/C
Palladium on carbon

Pd / C

What it's used for: Palladium adsorbed on charcoal (carbon) [Pd/C] is a heterogeneous catalyst. In the presence of hydrogen gas (H_2) it will convert alkenes and alkynes to alkanes, with *syn* addition of hydrogen.

Similar to: Lindlar's catalyst, "palladium" (Pd), platinum (Pt), platinum on carbon (Pt/C), nickel (Ni), ruthenium on carbon (Ru/C), rhodium on carbon (Rh)

Example 1: Reduction - conversion of alkenes to alkanes

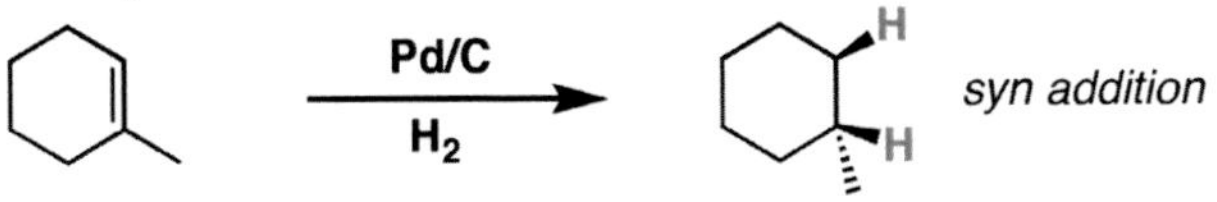

Pd/C, H_2 — H, H — *syn addition*

Example 2: Reduction - conversion of alkynes to alkanes

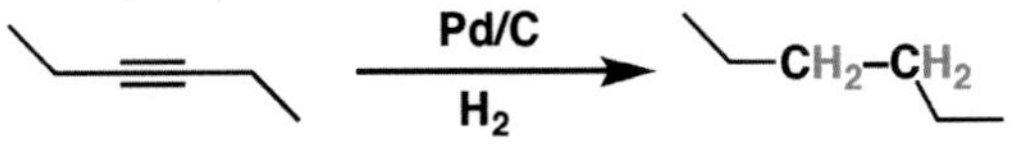

Pd/C, H_2 — CH_2–CH_2

Example 3: Reduction - conversion of nitro groups to primary amines

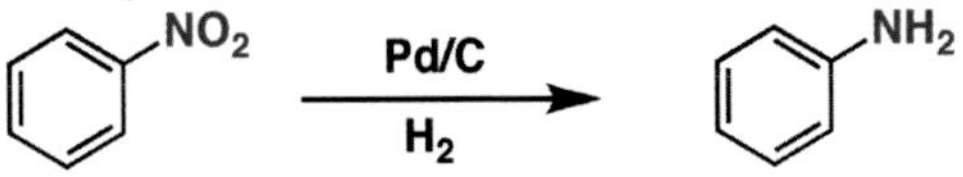

NO_2 → Pd/C, H_2 → NH_2

Example 4: Reduction - conversion of nitriles to primary amines

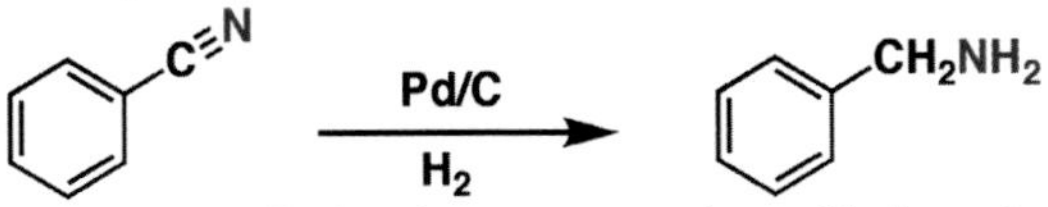

C≡N → Pd/C, H_2 → CH_2NH_2

Example 5: Reduction - conversion of imines to amines

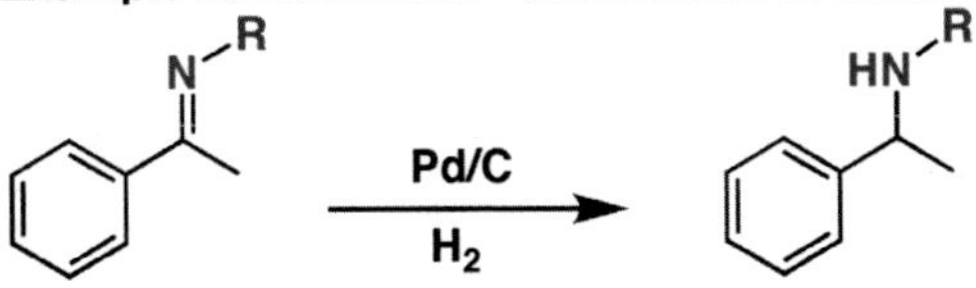

N–R → Pd/C, H_2 → HN–R

Example 6: Reduction - conversion of arenes to cyclic alkanes

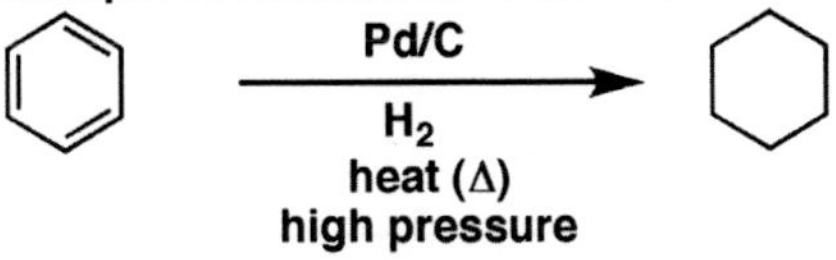

Pd/C, H_2, heat (Δ), high pressure

How it works: ***Hydrogenation***
Both hydrogen gas and the alkene are adsorbed onto the surface of the catalyst. The hydrogens are then delivered in *syn* fashion. Adsorbing palladium onto a material like charcoal (C) allows for a high surface area.

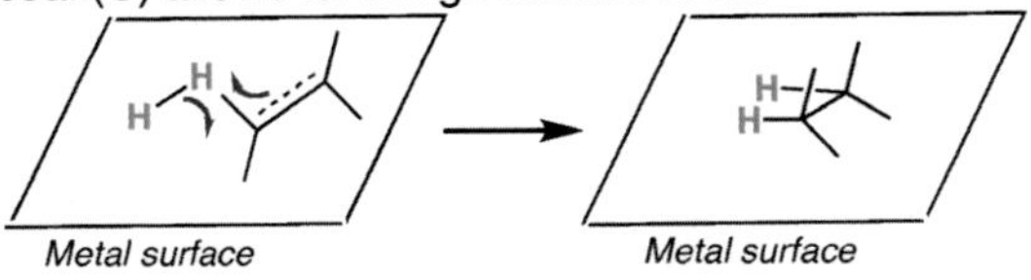

H, H → H, H

Metal surface → Metal surface

Pt
Platinum

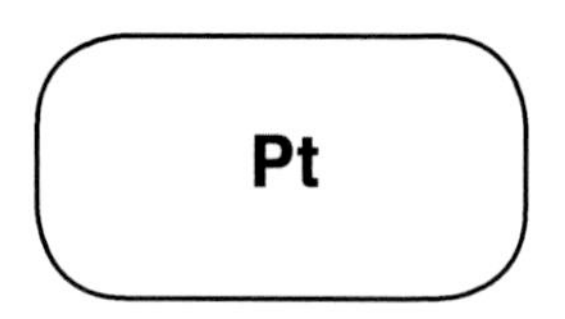

Also known as: Pt/C

What it's used for: Platinum, a "noble metal", is used for the reduction of carbon-carbon multiple bonds in the presence of hydrogen gas.

Similar to: Palladium on carbon (Pd/C), Nickel (Ni), Ruthenium on carbon (Ru/C)

Example 1: Reduction - conversion of alkenes to alkanes

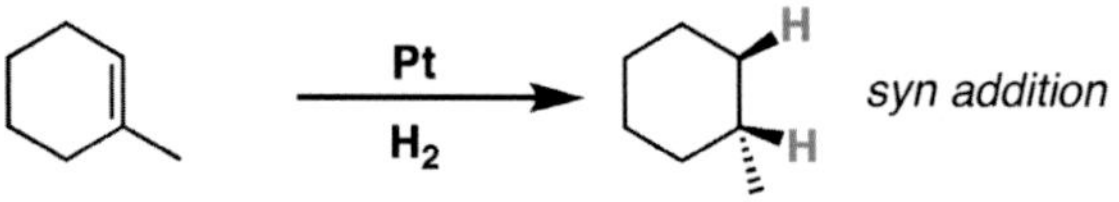

Pt, H_2 — *syn addition*

Example 2: Reduction - conversion of alkynes to alkanes

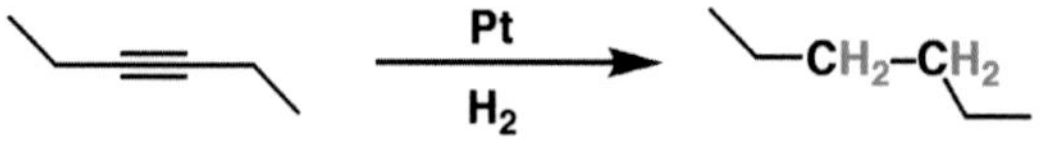

Pt, H_2 — CH_2–CH_2

Example 3: Reduction - conversion of arenes to cyclic alkanes

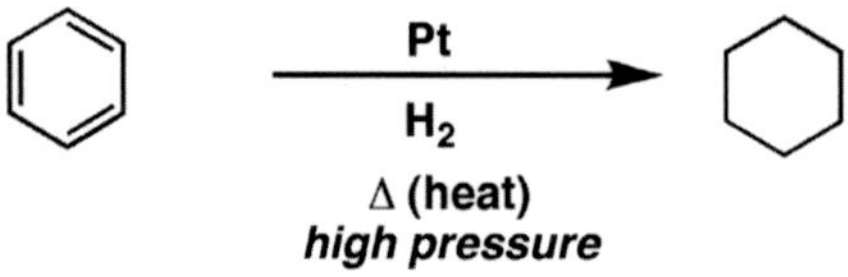

Pt, H_2, Δ (heat), ***high pressure***

How it works: ***Hydrogenation***

Like palladium on carbon (Pd/C), platinum is a heterogeneous catalyst. Both hydrogen gas and the alkene are adsorbed on to the surface of the catalyst. The hydrogens are then delivered in *syn* fashion.

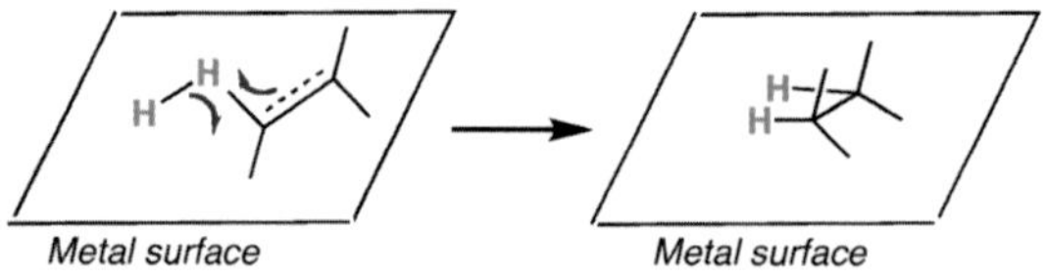

Metal surface → *Metal surface*

PCC
Pyridinium chlorochromate

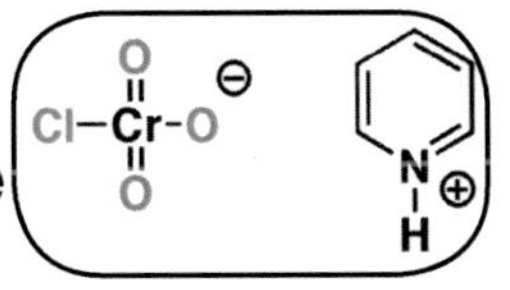

What it's used for: Pyridinium chlorochromate (PCC) is a reagent for the oxidation of primary alcohols to aldehydes and secondary alcohols. It is much milder than reactants such as H_2CrO_4 and $KMnO_4$ (which will oxidize primary alcohols to carboxylic acids).

Similar to: CrO_3/pyridine, Dess-Martin periodinane (DMP)

Example 1: Oxidation - conversion of primary alcohols to aldehydes

OH —PCC→ O, H

Example 2: Oxidation - conversion of secondary alcohols to ketones

OH —PCC→ O

How it works: ***Oxidation of primary alcohols to aldehydes***

The alcohol coordinates to the chromium (VI) atom, displacing chlorine, which then acts as a base, resulting in formation of a new C–O bond and reduction of Cr(VI) to Cr (IV).

Cl–Cr–O, O, O, OH, N–H — *Addition* → O, O, Cr, O, O, H, Cl, N–H

Proton transfer

O, OH, Cr, O, O, Cl, N–H — *Elimination of Cl* → :Cl:, H, H, O, OH, Cr, O, O

Elimination: formation of C–O bond

H, O + O, Cr, HO, OH **Cr(IV)** + N, H, Cl *Pyridinium chloride (byproduct)*

$POCl_3$
Phosphorus oxychloride

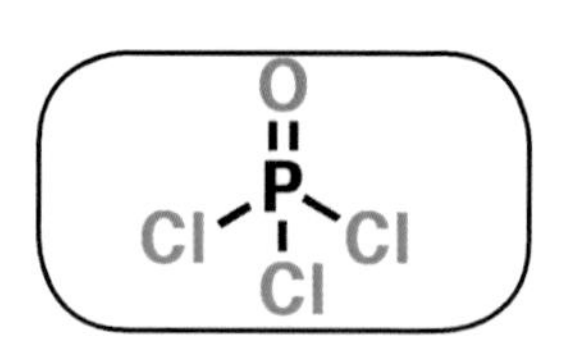

What it's used for: Phosphorus oxychloride ($POCl_3$) is used for the dehydration of alcohols to alkenes. Essentially it converts alcohols to a good leaving group, which is then removed by added base (often pyridine). It can also be used to convert amides to nitriles.

Similar to: $LiAlH_4$ (LAH), $LiAlH(Ot\text{-}Bu)_3$

Example 1: Elimination - conversion of alcohols to alkenes

HO

$POCl_3$

(pyridine)

pyridine is a base for this elimination reaction

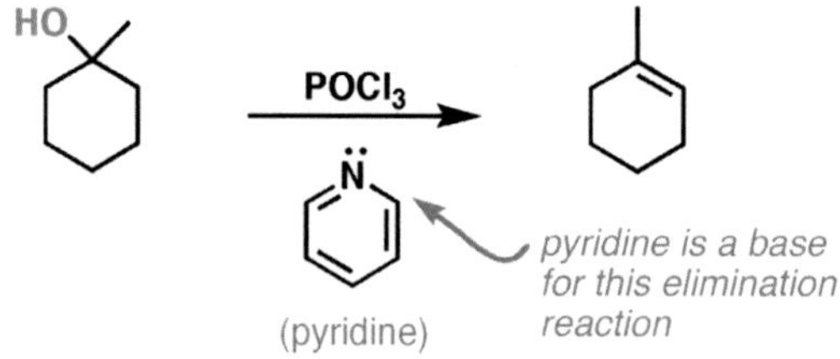

Example 2: Conversion of amides to nitriles

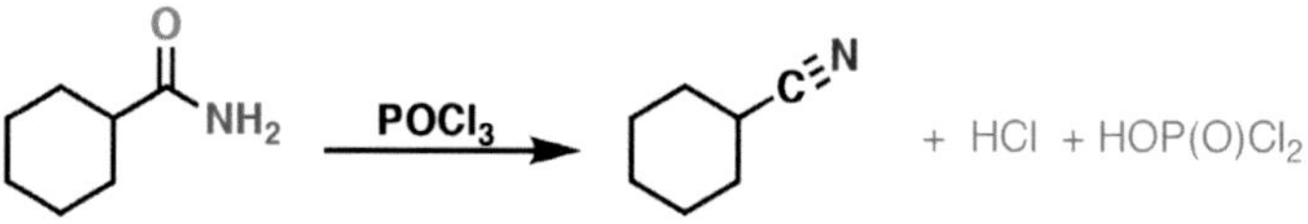

O
NH_2 $POCl_3$ C≡N + HCl + $HOP(O)Cl_2$

How it works: ***Elimination of alcohols to alkenes***

In this reaction, the alcohol oxygen attacks phosphorus, displacing chloride ion. Then, climination of the newly formed leaving group leads to formation of the alkene.

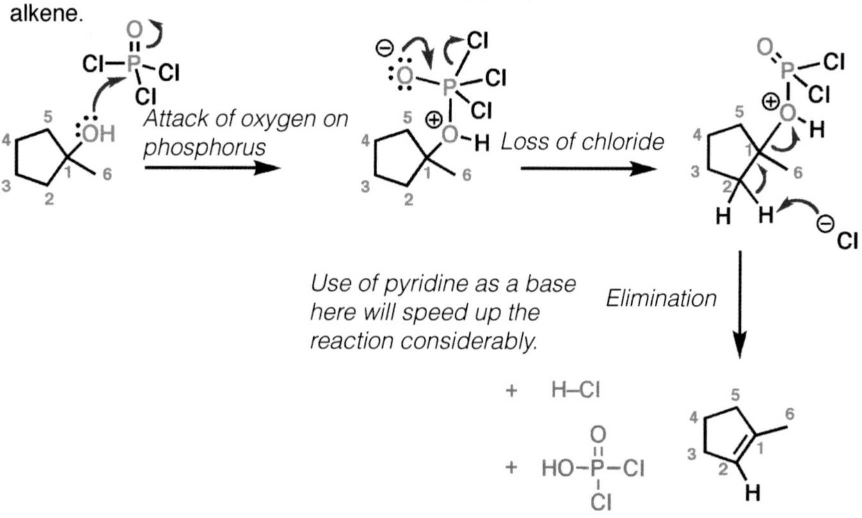

Attack of oxygen on phosphorus

Loss of chloride

Use of pyridine as a base here will speed up the reaction considerably.

Elimination

+ H–Cl

+ $HO-P(O)(Cl)-Cl$

PPh_3
Triphenylphosphine

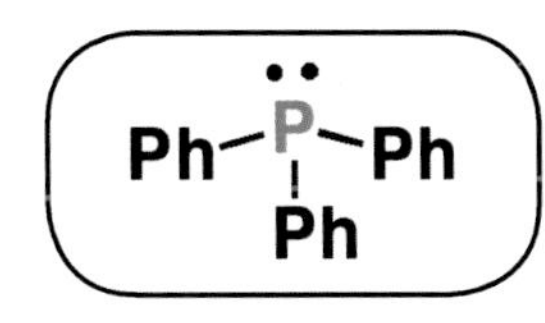

Also known as: Triphenylphosphane

What it's used for: PPh_3 is commonly used for formation of ylides in the Wittig reaction. It can also be used for reductive workup in the ozonolysis of alkenes.

Similar to: Dimethyl sulfide (in the reductive workup of ozonolysis)

Example 1: Wittig reaction - conversion of aldehydes/ketones to alkenes

Br
1) Ph_3P
2) *n*-BuLi, PhCHO
H
+ $Ph_3P=O$

How it works: ***The Wittig reaction***

Triphenylphosphine is a good nucleophile, and will react with alkyl halides to form phosphonium salts. Treatment of the phosphonium salt with a strong base such as *n*-BuLi results in the formation of an ylide. Reaction of the ylide with an aldehyde or ketone gives an oxaphosphatane, which opens to give an alkene and triphenylphosphine oxide.

:PPh_3
Br:
S_N2
Li⊕
⊖
H
⊕ Br⊖
PPh_3
Deprotonation with strong base

Phosphonium salt

Ph
H
O

Ph
O
⊖
PPh_3
⊕
+ LiBr +

Ylide

Ph
O
P
Ph_3
"oxaphosphatane"

Breakdown of the oxaphosphatane gives an alkene and triphenylphosphine oxide

Driving force for this reaction is formation of the strong P=O bond

Ph
+ $Ph_3P=O$
this is "triphenylphosphine oxide"

Pyridine

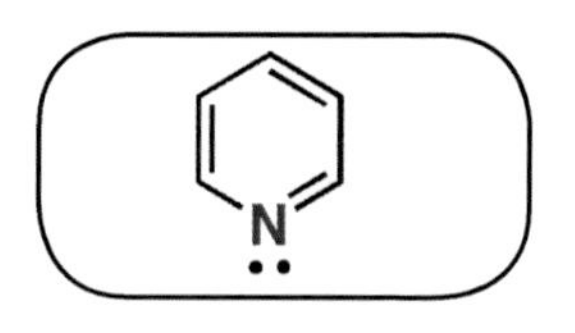

Also known as: Often abbreviated "pyr" or "py"

What it's used for: Pyridine is a mild base. Since it bears no charge it is especially soluble in organic solvents. It is often used in reactions that generate HCl and other strong acids - it acts much like a "sponge" for strong acid.

Similar to: Triethylamine (NEt_3), NaOH, other bases

Example 1: Conversion of alcohols to tosylates or mesylates

Pyridine acts as a weak base in this reaction, neutralizing the HCl generated here.

How it works: ***Formation of tosylates/mesylates***

Note that the role of pyridine here is simply to act as a base, removing the HCl generated during this reaction.

"Pyridinium chloride"

Ra–Ni
Raney Nickel

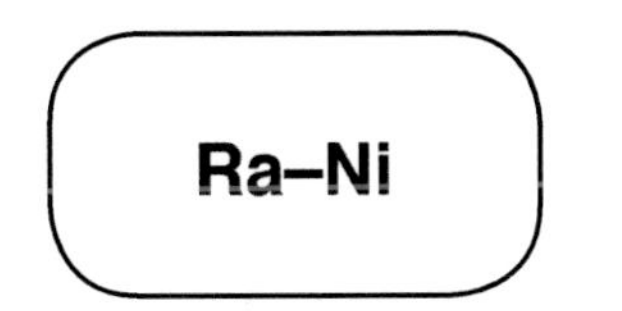

What it's used for: Raney nickel is a reagent for the reduction (hydrogenation) of double bonds. It is also used for the direct replacement of sulfur by hydrogen.

Similar to: Pd/C, Pt, Ni, and other heterogeneous catalysts.

Example 1: Reduction - conversion of thioacetals to alkanes

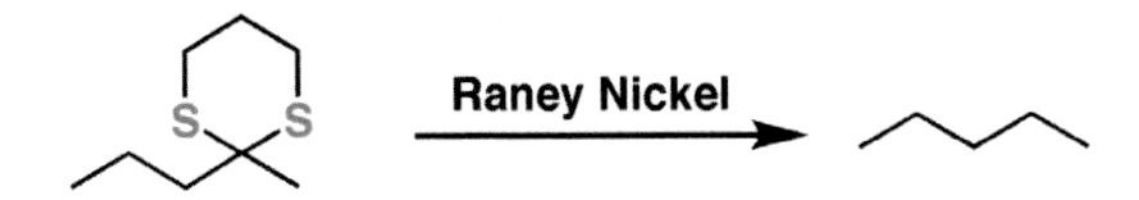

Example 2: Reduction - conversion of thioacetals to alkanes

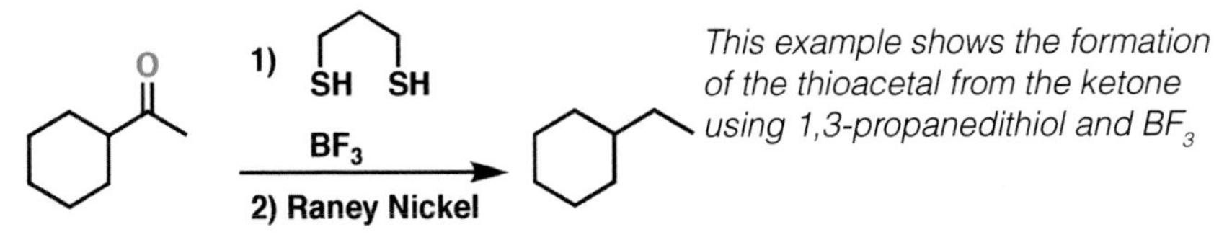

Example 3: Hydrogenation - conversion of alkenes to alkanes

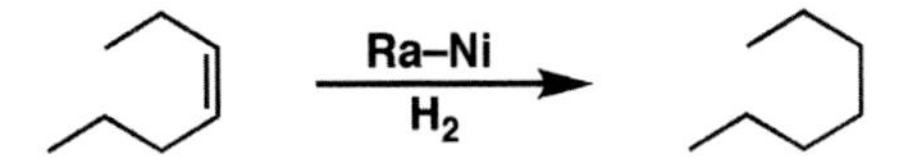

How it works: ***Reduction of thioacetals***

Raney Nickel is an alloy of nickel and aluminum which contains adsorbed hydrogen on its surface (hence, H_2 gas is not necessary in many cases!). Its exact mode of action for the reduction of C–S bonds is somewhat obscure.

RO–OR
Peroxides

What it's used for: Peroxides are used to initiate free-radical reactions. The oxygen-oxygen bond is very weak, and will fragment homolytically to generate free radicals.

Similar to: AIBN, benzoyl peroxide

Example 1: Free-radical addition - conversion of alkenes to alkyl bromides

HBr, RO–OR (peroxides)

Note that Br adds to the less substituted carbon "anti-Markovnikov"

How it works: ***Free radical bromination***

Peroxides, which have the general formula RO–OR, will fragment homolytically upon heating to provide alkoxy radicals RO•

RO—OR ⟶ 2 RO •

These radicals are very reactive and will readily remove hydrogen from various groups, leading to free radical chain processes.

RO • + H–Br ⟶ ROH + • Br

Addition occurs to the less substituted carbon (i.e. "anti-Markovnikov" because this results in the more stable (i.e. most substituted) secondary radical.

Propagation step 1

Propagation step 2

H–Br ⟶ + • Br

The bromine radical generated here can re-enter the reaction at propagation step 1 (addition to the alkene)

SO_3
Sulfur trioxide

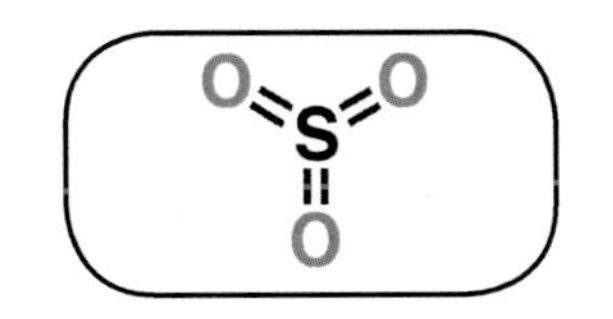

What it's used for: Sulfur trioxide is a reagent for the sulfonylation of aromatic groups. In the presence of acid, it will lead to the formation of sulfonic acids.

Example 1: Sulfonylation - conversion of arenes to aryl sulfonic acids

SO_3, H_2SO_4

How it works: ***Formation of sulfonic acids***

SO_3 is often used in the presence of sulfuric acid. Protonation of SO_3 gives SO_3H^+, a better electrophile than SO_3.

Often written HSO_4^-

The SO_3H^+ is then attacked by the aromatic ring:

Deprotonation regenerates aromaticity

$+ H_2SO_4$

Note that H_2SO_4 is regenerated in this step; it is a catalyst here.

$SOBr_2$
Thionyl bromide

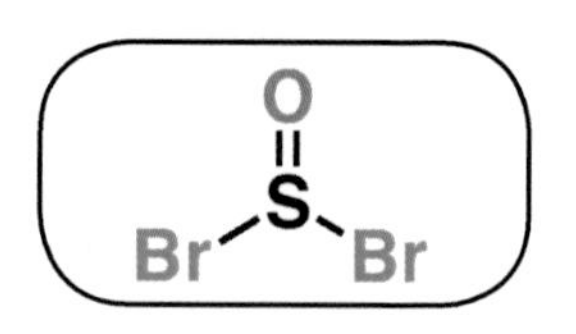

What it's used for: Thionyl bromide is a useful reagent for the formation of alkyl bromides from alcohols, as well as acid bromides (acyl bromides) from carboxylic acids.

Similar to: PBr_3, $SOCl_2$ (operates by the same mechanism)

Example 1: Conversion of alcohols to alkyl bromides

OH —$SOBr_2$→ Br + HBr + SO_2

note: inversion

Example 2: Conversion of carboxylic acids to acyl bromides

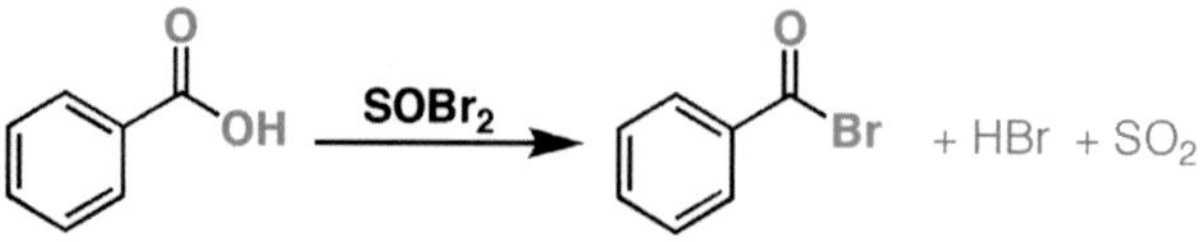

O, OH —$SOBr_2$→ O, Br + HBr + SO_2

How it works: ***Formation of alkyl bromides***

Identical in all respects to $SOCl_2$ (see section)

$SOCl_2$
Thionyl chloride

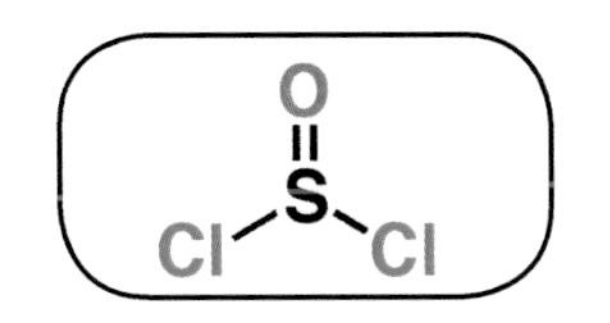

What it's used for: Thionyl chloride is used for the formation of alkyl chlorides from alcohols, and acid chlorides (acyl chlorides) from carboxylic acids.

Similar to: PCl_3, PCl_5, $SOBr_2$

Example 1: Conversion of alcohols to alkyl chlorides

OH — $SOCl_2$ → Cl + HCl + SO_2

note: inversion

Example 2: Conversion of carboxylic acids to acyl chlorides

O, OH — $SOCl_2$ → O, Cl + HCl + SO_2

How it works: ***Formation of alkyl chlorides***

This reaction proceeds through attack of oxygen on sulfur and then S_N2 attack of chloride ion on carbon, resulting in inversion of configuration. SO_2 gas is liberated.

+ HCl + SO_2

How it works: ***Formation of acyl chlorides***

Attack of oxygen at sulfur is followed by addition of chloride ion and elimination of O–SOCl, which loses Cl^- to become SO_2. The product is an acyl chloride.

Attack at sulfur

Elimination of chloride

Addition of chloride

Elimination of oxygen

Deprotonation

Acyl chloride

+ HCl + SO_2

Sn
Tin

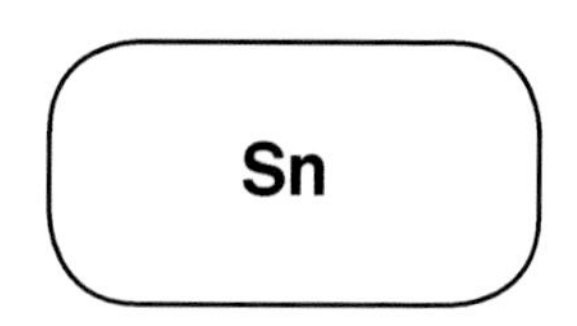

What it's used for: Tin is a reducing agent. In the presence of strong acids, it will reduce nitro groups to amines.

Similar to: Tin (II) chloride ($SnCl_2$), Iron metal (Fe), Zinc (Zn).

Example 1: Reduction - conversion of nitro groups to amines

$PhNO_2 \xrightarrow[HCl]{Sn} PhNH_2 + SnCl_2 + H_2O$

How it works: ***Reduction of nitro groups***

The mechanism has not been determined with 100% certainty. Often not depicted in textbooks. Here is a possible mechanism.

Sn donates electron pair

loss of water

$SnCl_2$

+ :SnCl

Sn donates electron pair to N=O (nitroso)

+ H_2O

+ $SnCl_2$

Hydroxylamine

loss of water

Sn donates electron pair

Amine

+ $SnCl_2$

TBAF

Tetrabutyl ammonium fluoride

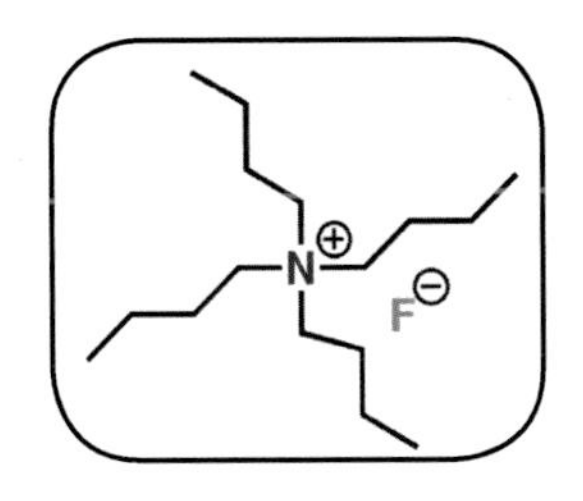

What it's used for: Tetrabutyl ammonium fluoride (TBAF) is a source of fluoride ion. It is used to cleave silyl ethers, which are common protecting groups for alcohols (fluorine forms very strong bonds with silicon)

Example 1: Alcohol deprotection - conversion of silyl ethers to alcohols

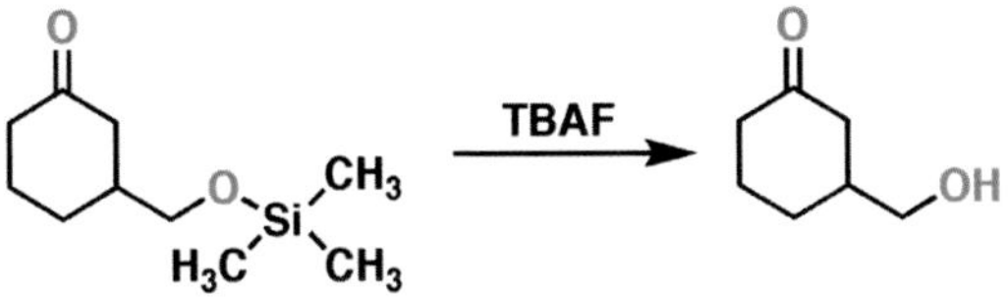

TBAF

How it works: ***Deprotection of silyl ethers***

Fluorine forms very strong bonds to silicon. It is on this principle that fluoride ion acts to break O-Si bonds and form Si-F bonds. Since TBAF is generally used as a solution in water, the resulting alkoxide is protonated by the aqueous solvent.

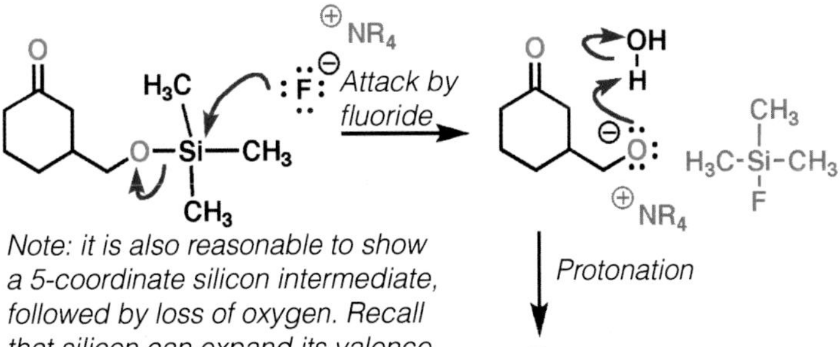

Attack by fluoride

Note: it is also reasonable to show a 5-coordinate silicon intermediate, followed by loss of oxygen. Recall that silicon can expand its valence shell.

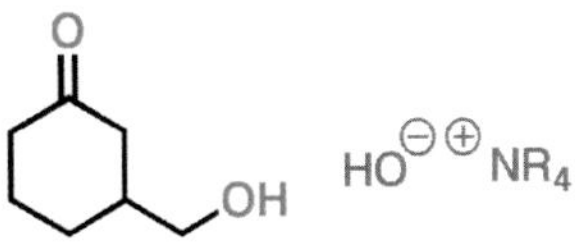

Protonation

Note that the tetrabutylammonium cation here is abbreviated NR_4^+

TMSCl

Trimethylsilyl chloride

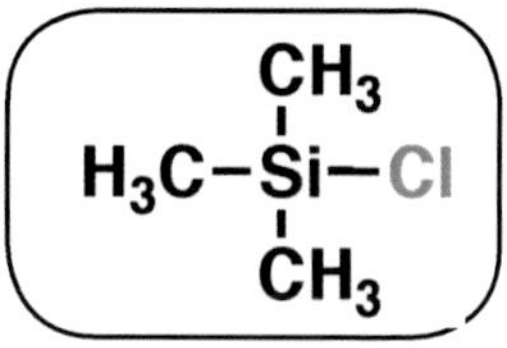

$H_3C-Si(CH_3)_2-Cl$

Also known as: Chlorotrimethylsilane, $(CH_3)_3SiCl$

What it's used for: TMSCl i s a protecting group for alcohols. When added to alcohols it is inert to most reagents except for fluoride ion (F^-) and acid. The addition of a weak base such as pyridine can serve to remove the HCl byproduct that is formed during this reaction.

Similar to: TBSCl

Example 1: Alcohol protection - conversion of alcohols to silyl ethers

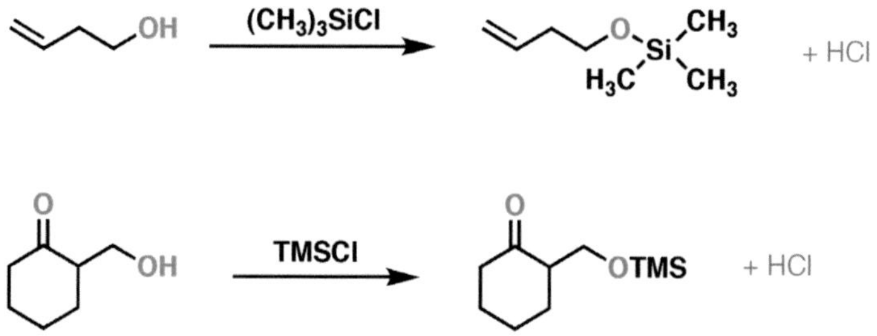

$(CH_3)_3SiCl$ + HCl

TMSCl OTMS + HCl

How it works: ***Protection of alcohols***

The reaction for protection of alcohols is quite straightforward - attack of the alcohol on silicon accompanied by loss of the chloride leaving group. Addition of a base such as pyridine will neutralize the HCl that is formed during the reaction.

TsCl
p-Toluenesulfonyl chloride

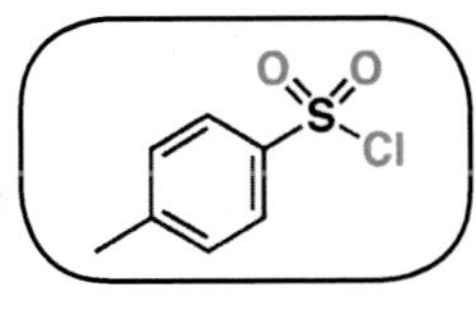

Also known as: TosCl, p-TsCl, Tosyl choride

What it's used for: Tosyl chloride (TsCl) will convert alcohols to sulfonates, which are excellent leaving groups in elimination and substitution reactions. TsO^- is the conjugate base of the strong acid TsOH.

Similar to: Mesyl chloride (MsCl), p-bromobenzenesulfonyl chloride (BsCl)

Example 1: Conversion of alcohols to alkyl tosylates

OH → TsCl, pyridine → OTs + pyridinium $H^{\oplus}$ $Cl^{\ominus}$

pyridine (a weak base) reacts with the HCl that is generated

Example 2: Substitution of tosylates

TsO → NaCN, DMSO → N≡C + NaOTs

a polar aprotic solvent such as DMSO will increase the rate of this S_N2 reaction.

Example 3: Elimination of tosylates

OTs → NaOEt, Δ (heat) → + EtOH + NaOTs

How it works: ***Tosylates as leaving groups***

Weak bases are excellent leaving groups. By converting OH (a strong base and poor leaving group) to OTs (a much weaker base and leaving group) it becomes many orders of magnitude easier to do substitution and elimination reactions.

:Ö: :Ö: S :Ö: ⊖

Note how the negative charge on oxygen can be delocalized on to the other oxygen atoms through resonance.

TsOH
p-Toluenesulfonic acid

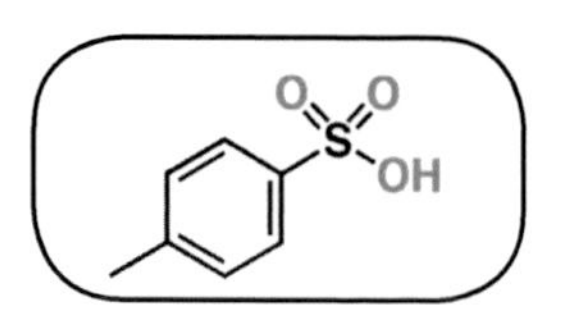

Also known as: Tosic acid, TosOH,

What it's used for: STosic acid is a strong acid, similar in strength to sulfuric acid (pka of –2.8). One feature is that the conjugate base is a poor nucleophle, which makes it useful for the dehydration of alcohols to form alkenes. Also it is a white crystalline solid, which makes it slightly more convenient to use than H_2SO_4 in some cases.
Similar to: Sulfuric acid (H_2SO_4)

Example 1: Elimination - conversion of alcohols to alkenes

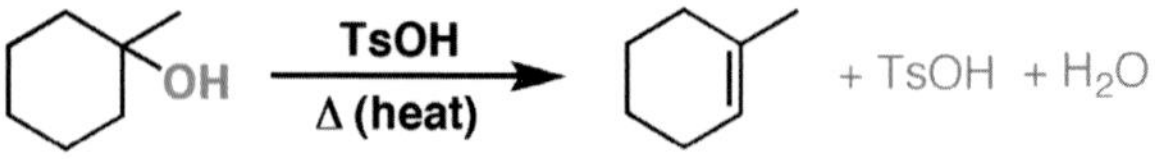

OH — TsOH, Δ (heat) → + TsOH + H_2O

How it works: ***Acid-catalyzed elimination of alcohols to give alkenes***

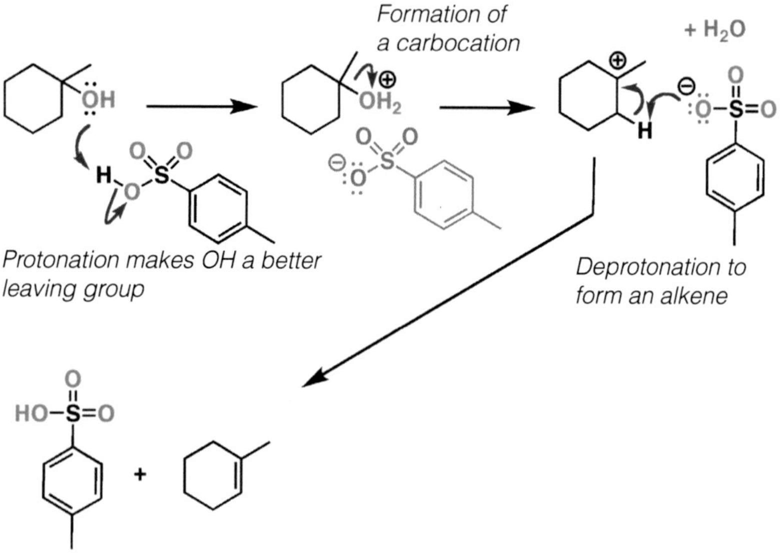

Formation of a carbocation

+ H_2O

Protonation makes OH a better leaving group

Deprotonation to form an alkene

HO–S=O
+

Note that TsOH is regenerated here (it is a catalyst in this reaction)

Zn
Zinc

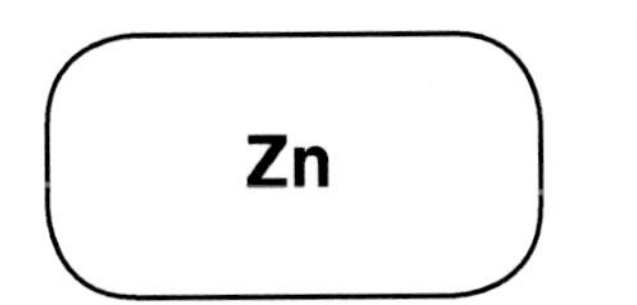

What it's used for: Zinc is a metal reducing agent. It is useful for the reduction of ozonides, and also in the reduction of nitro groups to amines (in the presence of acid)
Similar to: Dimethyl sulfide (in workup of ozonolysis), Sn (in reduction of nitro groups)

Example 1: Ozonolysis (reductive workup) - conversion of alkenes to aldehydes/ketones

1) O_3
2) Zn

Example 2: Reduction - conversion of nitro groups to primary amines

NO_2 → Zn, HCl → NH_2

How it works: ***Reduction of ozonides***

Zinc is easily oxidized and can donate electrons to various groups. One application is in the reduction of ozonides.

Ozonide

+ ZnO

For a suggestion on how metals such as Zn, Sn, and Fe reduce nitro groups in the presence of acids such as HCl, see the section on tin (Sn)

Zn/Cu
Zinc-Copper Couple

Zn(Cu)

What it's used for: Zinc-copper couple is a reducing agent, used to form carbenes (actually "carbenoids") from alkyl dihalides. When these are added to alkenes, they form cyclopropanes.

Example 1: Simmons-Smith reaction - formation of cyclopropanes from alkenes

Zn(Cu)
CH_2I_2
H
H

note: product is always syn

How it works: ***Cyclopropanation of alkenes***

Zinc-copper couple is an alloy of zinc and copper that can reduce dihalides to metal carbenoids. The carbenoid performs the cyclopropanation reaction.

I I Zn(Cu) I ZnI R R + ZnI_2

"carbenoid" ***cyclopropane***

Zinc-copper couple itself is a metal cluster (not a molecule). The mechanism of action for formation of the Zn-carbon bond is probably similar to that of organolithium and Grignard reagents.

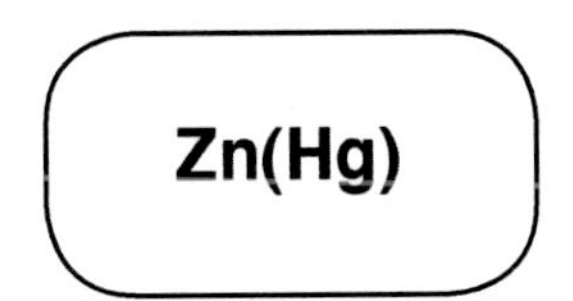

Zinc amalgam

What it's used for: In the presence of acid, zinc amalgam will reduce ketones to alkanes, in a process called the Clemmensen reaction (or Clemmensen reduction)

Example 1: The Clemmensen Reduction

O — Zn(Hg) / HCl → H H

Note: these reactions tend to work best on ketones adjacent to aromatic rings.

How it works: ***The Clemmensen Reduction***

Metals that are fused with mercury are called *amalgams*, and their precise mode of action is somewhat obscure. The strong acid (HCl) serves to activate the carbonyl toward reduction, as well as assist in the eventual removal of the carbonyl oxygen as H_2O

Odds And Ends

DBU (1,8-Diazabicycloundec7-ene)

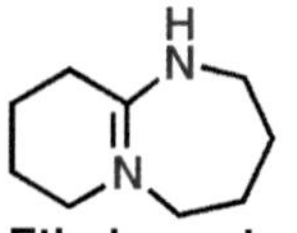

Base; used in elimination reactions. Bulky, non-nucleophilic base.

Ethylene glycol

Used as solvent in Wolff Kishner and as a protecting group (acetal) for ketones. High boiling solvent.

$KHSO_3$ (Potassium bisulfate) Used in dihydroxylation reaction workup; breaks up "osmate ester".

NMO (N-methylmorpholine N-oxide)

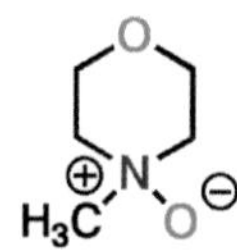

Used as oxidant in dihydroxylation reaction of alkenes. Allows for catalytic use of (expensive) OsO_4

Oxalyl chloride

Simlar to thionyl chloride; used in Swern oxidation of alcohols to give aldehydes and ketones. Converts DMSO to active electrophile.

See a reagent in your introductory course or textbook that isn't depicted anywhere in the Guide?

Click To Leave Feedback

Common Abbreviations and Terms

Abbreviation	Formula	Name
Me	$-CH_3$	*Methyl*
Et	$-CH_2CH_3$	*Ethyl*
Pr	$-CH_2CH_2CH_3$	*Propyl*
Bu	$-CH_2CH_2CH_2CH_3$	*Butyl*
***i*-Pr**	CH_3CHCH_3	*isopropyl*
***s*-Bu**	$CH_3CHCH_2CH_3$	*sec-butyl*
***i*-Bu**	$CH_2CH_2CH_3$, CH_3	*isobutyl*
***t*-Bu**	CH_3CCH_3, CH_3	*tert-butyl*
Ph (C_6H_5–)	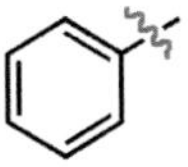	*phenyl*
Bn ($C_6H_5CH_2$–)	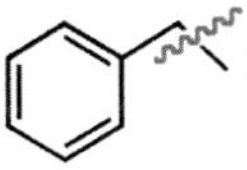	*benzyl*

Abbreviation	Name
Ac	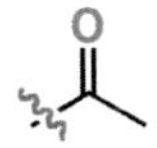 *acetyl*
Ts	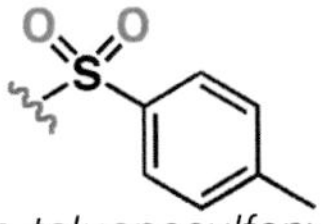 *p-toluenesulfonyl (tosyl)*
Ms	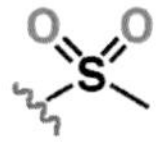 *methanesulfonyl (mesyl)*
Bs	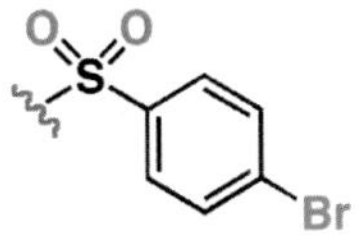 *p-bromobenzenesulfonyl (brosyl)*
Allyl	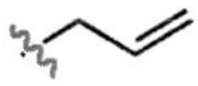
Vinyl	
Propargyl	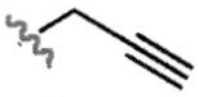
Piv	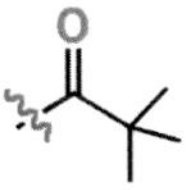 *Pivaolyl*

1°	**primary**
2°	**secondary**
LG	**leaving group**
Nu:	**nucleophile**
B:	**base**
R	**any carbon substituent**
Ar	**an aromatic substituent**
3°	**tertiary**
4°	**quaternary**
***N*-**	**denotes a substituent directly bound to nitrogen**
EDG	**electron donating group**
EWG	**electron withdrawing group**
Acyl	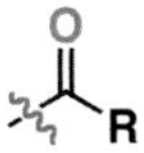
Carbonyl	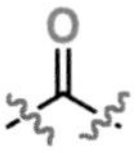

Functional Groups

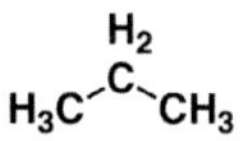

$H_3C-CH_2-CH_3$ alkane	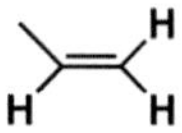 alkene	 alkyne	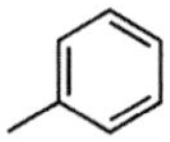 benzene ring (phenyl)	R–F: R–Cl: R–Br: R–I: alkyl halide
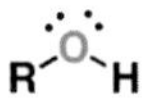 alcohol	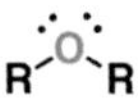 ether	 hemiacetal	 acetal	 epoxide
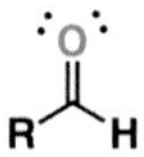 aldehyde	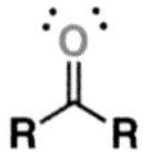 ketone	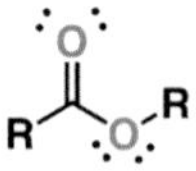 ester	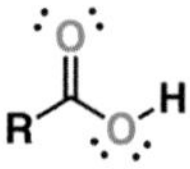 carboxylic acid	$N(H,R)_2$ amide
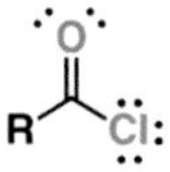 acid chloride	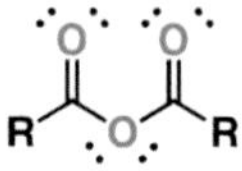 anhydride	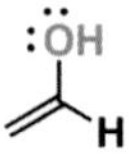 enol	$R-N(H, R)_2$ amine	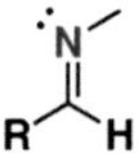 imine
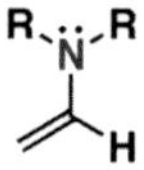 enamine	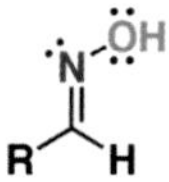 oxime	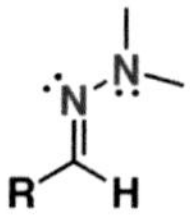 hydrazone	R–C≡N: nitrile	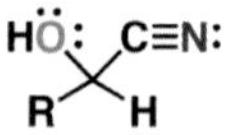 cyanohydrin
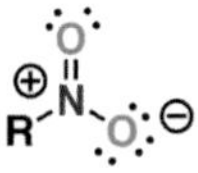 nitro	 sulfide (thioether)	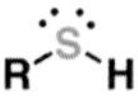 thiol	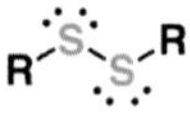 disulfide	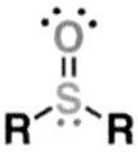 sulfoxide
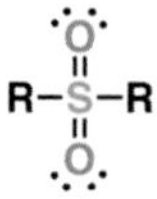 sulfone	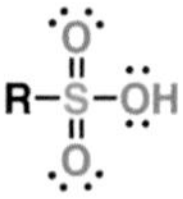 sulfonic acid	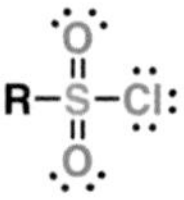 sulfonyl chloride	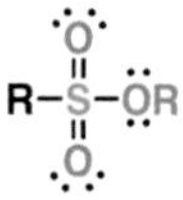 sulfonate ester	

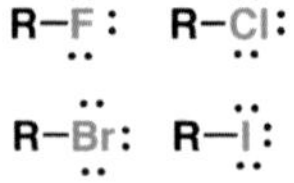

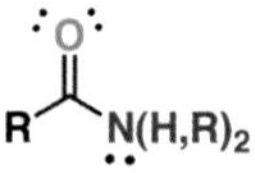

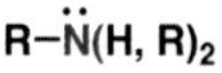

pKas of Common Functional Groups

Functional group	Example	pKa	Conjugate Base
Hydroiodic acid	HI	–10	$I^{\ominus}$
Hydrobromic acid	HBr	–9	$Br^{\ominus}$
Hydrochloric acid	HCl	–6	$Cl^{\ominus}$
Sulfuric acid	H_2SO_4	–3	$HSO_4^{\ominus}$
Sulfonic acids	Me–C6H4–S(=O)2–OH (tosic acid)	–3	Me–C6H4–S(=O)2–$O^{\ominus}$
Hydronium ion	$H_3O^{\oplus}$	–1.7	H_2O
Hydrofluoric acid	H–F	3.2	$F^{\ominus}$
Carboxylic acids	H_3C–C(=O)–OH	~4	H_3C–C(=O)–$O^{\ominus}$
Protonated amines	$NH_4^{\oplus}$ $Cl^{\ominus}$	9-11	NH_3
Thiols	CH_3S–H	13	$CH_3S^{\ominus}$
Malonates	H_3CO–C(=O)–CH(H)(H)–C(=O)–OCH_3	13	H_3CO–C(=O)–$C^{\ominus}$(H)–C(=O)–OCH_3
Water	HO–H	16	$HO^{\ominus}$

pKas of Common Functional Groups

Functional group	Example	pKa	Conjugate Base
Alcohol	$H_3C{-}OH$	17	$H_3C{-}O^{\ominus}$
Ketone/ aldehyde	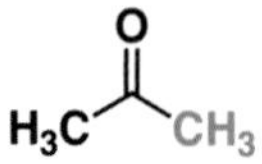 $H_3C{-}C(=O){-}CH_3$	20-24	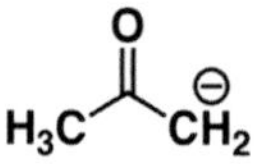 $H_3C{-}C(=O){-}CH_2^{\ominus}$
Ester	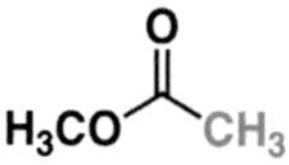 $H_3CO{-}C(=O){-}CH_3$	25	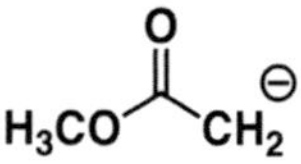 $H_3CO{-}C(=O){-}CH_2^{\ominus}$
Nitrile	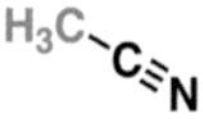 $H_3C{-}C{\equiv}N$	25	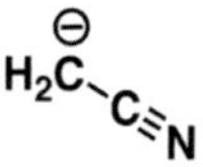 $H_2C^{\ominus}{-}C{\equiv}N$
Alkyne	$R{-}C{\equiv}C{-}H$	25	$R{-}C{\equiv}C^{\ominus}$
Sulfoxide	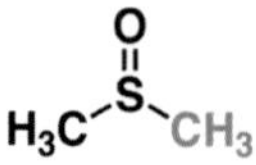 $H_3C{-}S(=O){-}CH_3$	31	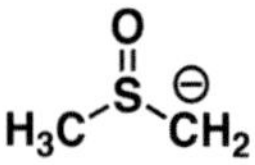 $H_3C{-}S(=O){-}CH_2^{\ominus}$
Amine	NH_3	~35-38	$^{\ominus}NH_2$
Hydrogen	H–H	42	$H^{\ominus}$
Alkene	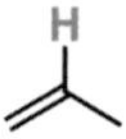 $CH_2{=}C(H){-}CH_3$	~43	 $CH_2{=}C^{\ominus}{-}CH_3$
Alkane	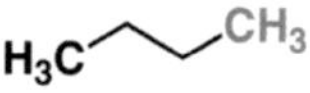 $H_3C{-}CH_2{-}CH_2{-}CH_3$	~50	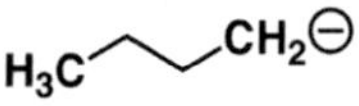 $H_3C{-}CH_2{-}CH_2{-}CH_2^{\ominus}$

http://masterorganicchemistry.com

Notes on Acids

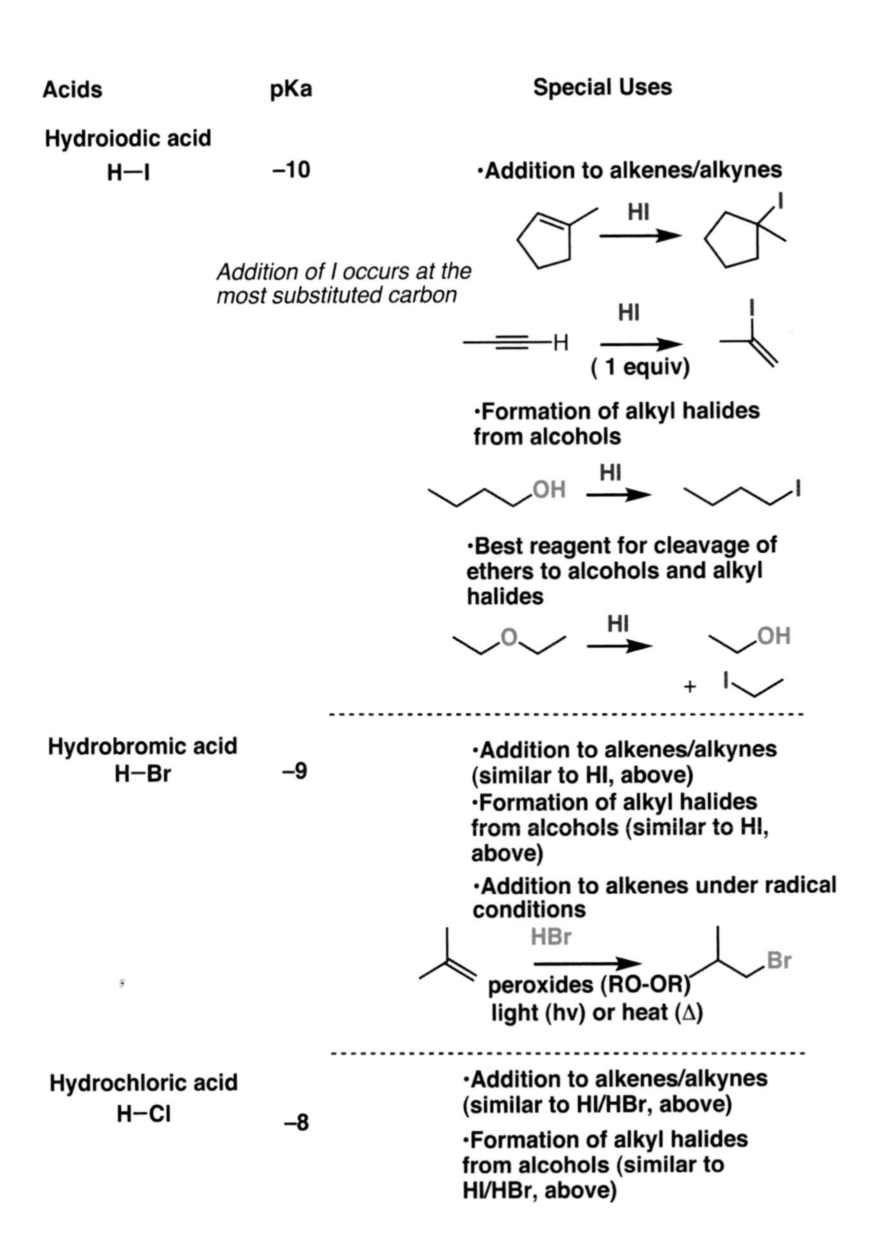

Acids	pKa	Special Uses
Hydroiodic acid H—I	–10	•Addition to alkenes/alkynes *Addition of I occurs at the most substituted carbon* HI HI (1 equiv) •Formation of alkyl halides from alcohols HI •Best reagent for cleavage of ethers to alcohols and alkyl halides HI
Hydrobromic acid H–Br	–9	•Addition to alkenes/alkynes (similar to HI, above) •Formation of alkyl halides from alcohols (similar to HI, above) •Addition to alkenes under radical conditions HBr peroxides (RO-OR) light (hv) or heat (Δ)
Hydrochloric acid H–Cl	–8	•Addition to alkenes/alkynes (similar to HI/HBr, above) •Formation of alkyl halides from alcohols (similar to HI/HBr, above)

Notes on Acids

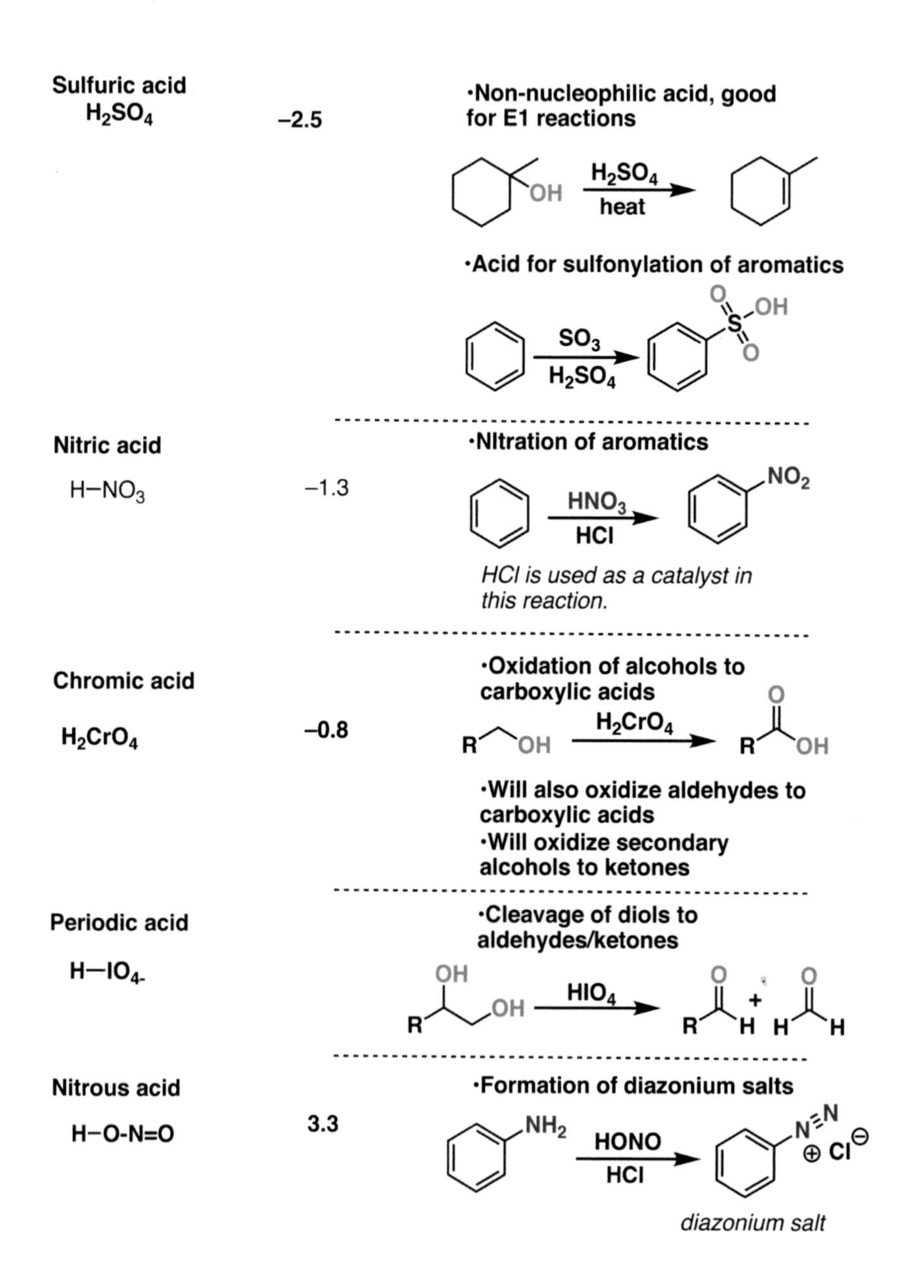

Sulfuric acid
H₂SO₄
–2.5
·Non-nucleophilic acid, good for E1 reactions
OH
H₂SO₄
heat
·Acid for sulfonylation of aromatics
SO₃
H₂SO₄
O
S
OH
O
Nitric acid
H–NO₃
–1.3
·NItration of aromatics
HNO₃
HCl
NO₂
HCl is used as a catalyst in this reaction.
Chromic acid
H₂CrO₄
–0.8
·Oxidation of alcohols to carboxylic acids
R
OH
H₂CrO₄
O
R
OH
·Will also oxidize aldehydes to carboxylic acids
·Will oxidize secondary alcohols to ketones
Periodic acid
H—IO₄-
·Cleavage of diols to aldehydes/ketones
OH
R
OH
HIO₄
O
R
H
+
O
H
H
Nitrous acid
H–O-N=O
3.3
·Formation of diazonium salts
NH₂
HONO
HCl
N
N
⊕ Cl⊖
diazonium salt

Notes on Bases

Base	pKa (of conjugate acid)	Use
Pyridine 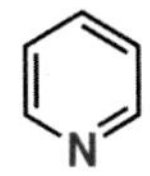	5	• Soluble in organic solvents, useful for reactions that generate HCl and HBr, such as: 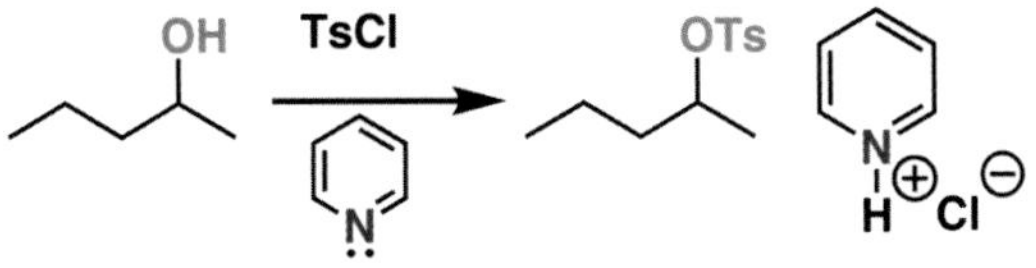 OH → (TsCl, pyridine) → OTs + pyridinium (N^+–H) Cl^-
Sodium tert-butoxide 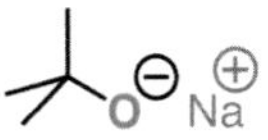	19	• Strong, bulky base for elimination reactions, tends to give less-substituted alkene 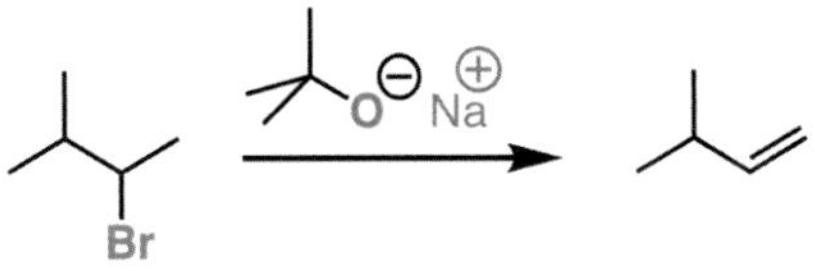 (Br, O^- Na^+)
LDA (Li^+ N^-) 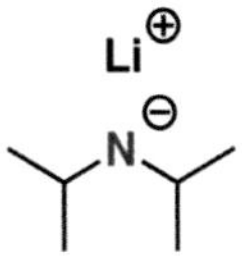	36	• Formation of less-substituted enolates:  O → (LDA, THF, –78°C) → OLi
Sodium amide (Na^+ NH_2^-)	38	• Elimination of dihalides to give alkynes: 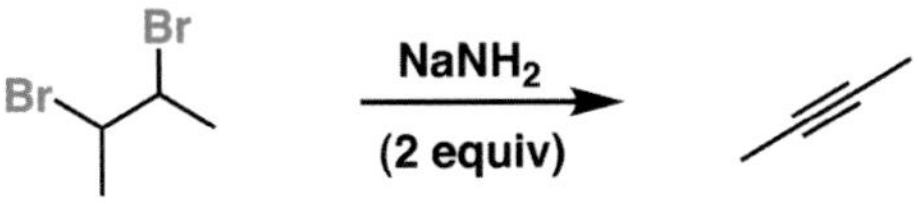 Br, Br → ($NaNH_2$ (2 equiv)) → alkyne • Deprotonation of alkynes: 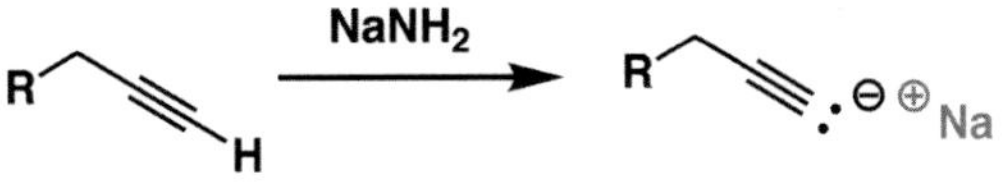 R–C≡C–H → ($NaNH_2$) → R–C≡C:$^-$ Na^+
Sodium hydride (NaH)	42	• Strong, non-nucleophilic base: 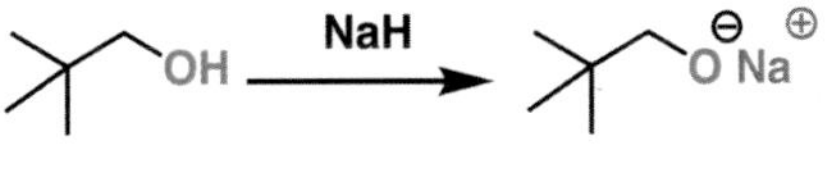 OH → (NaH) → O^- Na^+

Oxidizing Agents

Transformation		Reagent
Primary Alcohol	Aldehyde	• PCC • CrO_3 / pyridine
Secondary Alcohol	Ketone	• PCC • CrO_3 / pyridine • H_2CrO_4 (note: same as $K_2Cr_2O_7/H_2SO_4$ or $Na_2Cr_2O_7/H_2SO_4$, or CrO_3/H+) • $KMnO_4$
Aldehyde	Carboxylic acid	• H_2CrO_4 (see note above) • $KMnO_4$ • H_2O_2
Alcohol	Carboxylic acid	• $KMnO_4$ • H_2CrO_4 (see note above)
Alkane	Carboxylic acid	• $KMnO_4$
Alkene	Aldehyde / Ketone	• O_3 , then Zn • O_3 , then CH_3SCH_3 (DMS)
Alkene	Carboxylic acids/ Ketones	• O_3 , then H_2O_2 • $KMnO_4$, heat, $H_3O^{\oplus}$
Alkyne	Carboxylic acids	• O_3 , then H_2SO_4 • $KMnO_4$, heat, $H_3O^{\oplus}$

Oxidizing Agents

Transformation		Reagent
Alkene	Diol (vicinal diol) HO OH *always syn*	• OsO_4 • $KMnO_4$, $HO^{\ominus}$
Alkene	Epoxide O	• mCPBA
Diol HO OH	Aldehyde O H	• $NaIO_4$ • $Pb(OAc)_4$ • HIO_4
Ketone O	Ester O O	• mCPBA

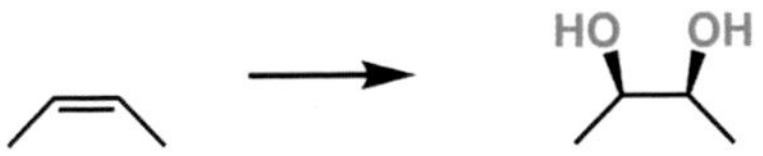

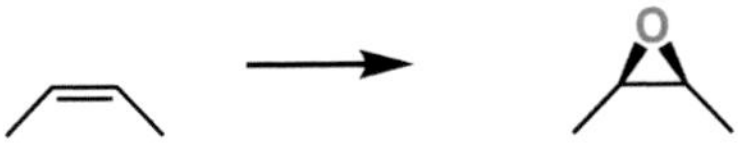

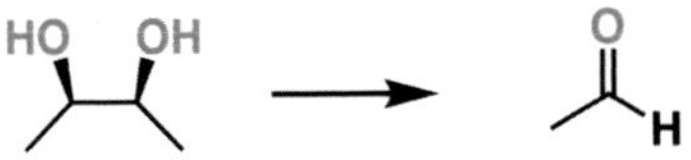

Reducing Agents

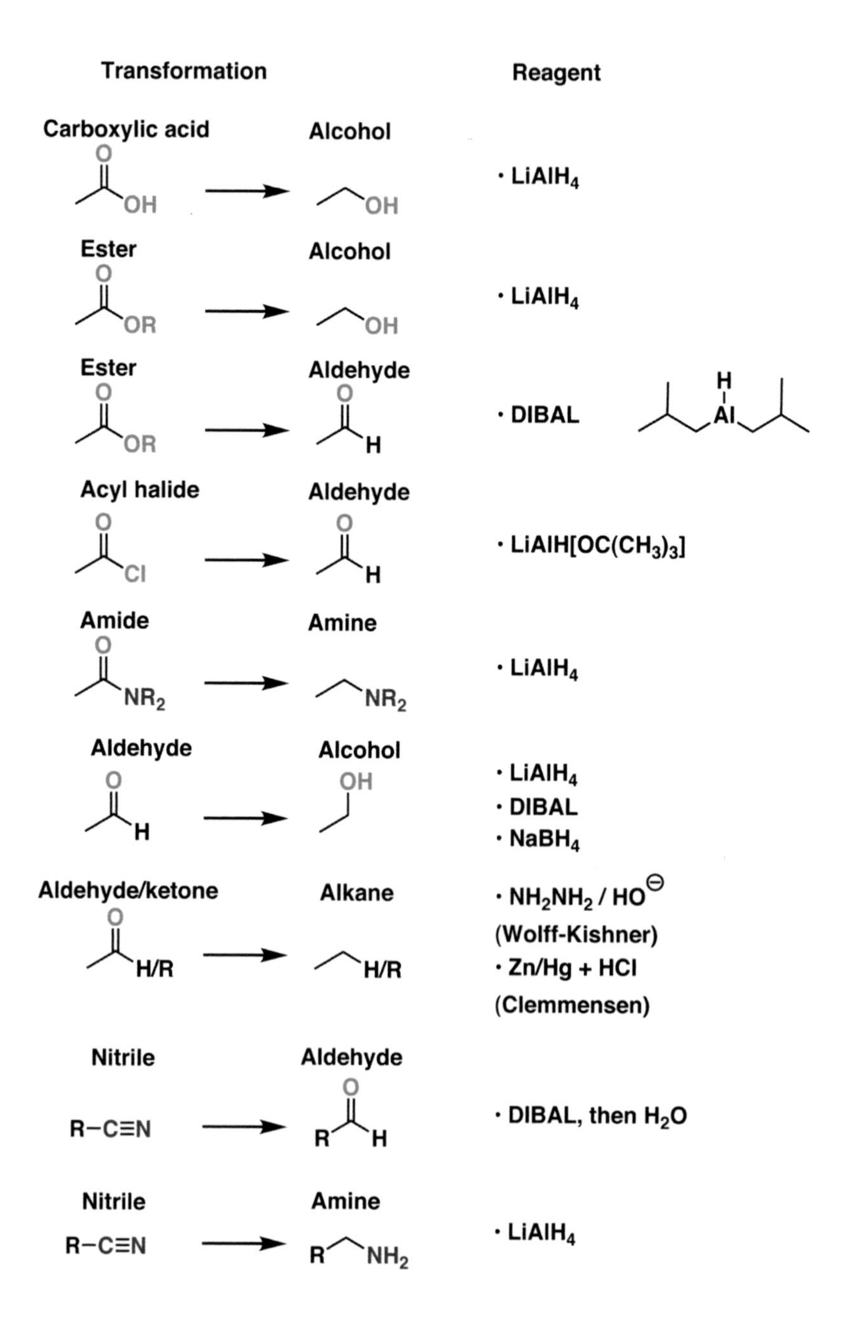

Reducing Agents

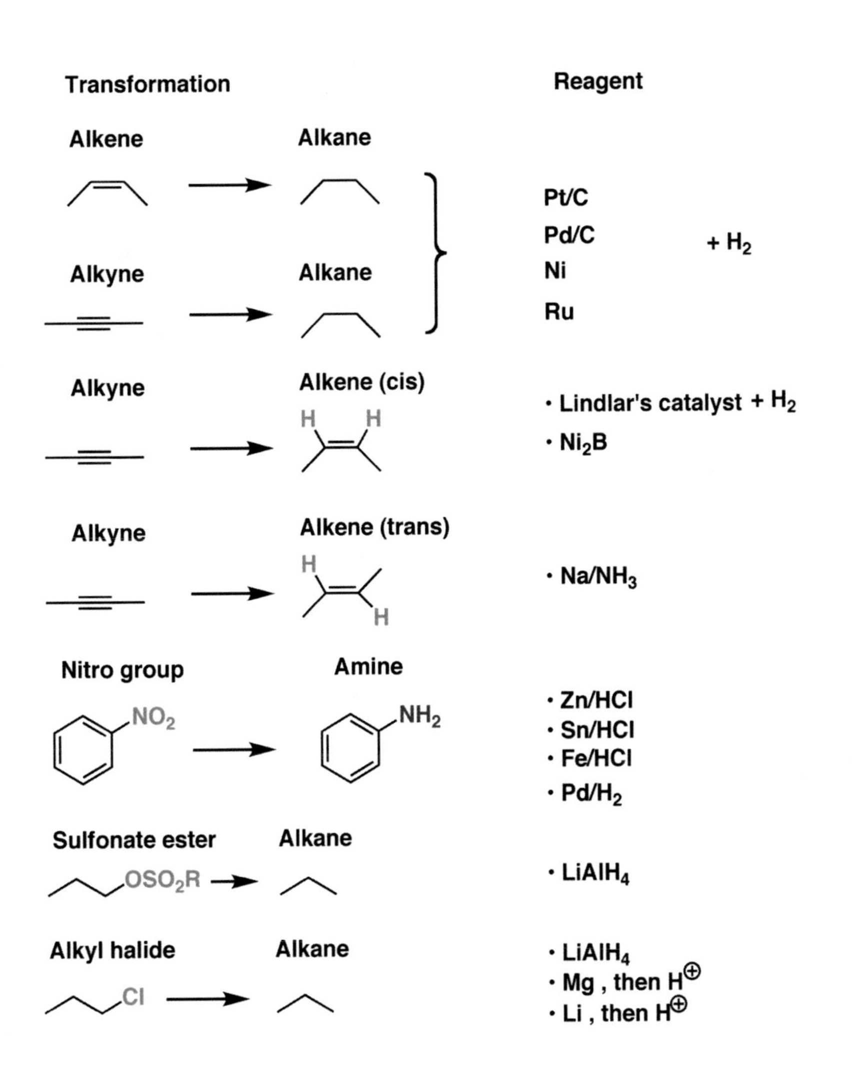
Transformation
Reagent
Alkene
Alkane
Alkyne
Alkane
Pt/C
Pd/C
Ni
Ru
+ H2
Alkyne
Alkene (cis)
H
H
• Lindlar's catalyst + H2
• Ni2B
Alkyne
Alkene (trans)
H
H
• Na/NH3
Nitro group
Amine
NO2
NH2
• Zn/HCl
• Sn/HCl
• Fe/HCl
• Pd/H2
Sulfonate ester
Alkane
OSO2R
• LiAlH4
Alkyl halide
Alkane
Cl
• LiAlH4
• Mg , then H⊕
• Li , then H⊕

Organometallic Reagents

Conversion of aldehydes to secondary alcohols

R–CHO ⟶ R(H)C(OH)–R

• R–MgX (Grignard Reagents)
• R–Li (Organolithium reagents)

Conversion of ketones to tertiary alcohols

R–CO–R ⟶ R_3C–OH

• R–MgX
• R–Li

Conversion of acyl halides to tertiary alcohols

R–CO–Cl ⟶ R_3C–OH

• R–MgX
• R–Li

Conversion of acyl halides to ketones

R–CO–Cl ⟶ R–CO–R

• R_2CuLi (Organocuprate reagents)

Conversion of anhydrides to tertiary alcohols

R–CO–O–CO–R ⟶ R_3C–OH

• R–MgX
• R–Li

Conversion of esters to tertiary alcohols

R–CO–OR ⟶ R_3C–OH

• R–MgX
• R–Li

Conversion of carboxylic acids to ketones

R–CO–OH ⟶ R–CO–R

• R–Li

Conversion of nitriles to ketones

R–C≡N ⟶ R–CO–R

• R–MgX
• R–Li

reaction proceeds through an imine

Organometallic Reagents

Opening of epoxides

R-epoxide → R–CH(OH)–CH₂–R

- R–MgX
- R–Li

Reaction with alkyl halides or tosylates

allyl–Br → allyl–R

R–Br → R–R

- R_2CuLi

Conjugate addition (1,4 addition)

methyl vinyl ketone → R–CH₂CH₂–C(=O)–CH₃

- R_2CuLi

Addition to carbon dioxide to form carboxylic acids

O=C=O → HO–C(=O)–R

- R–MgX
- R–Li

Reaction with acidic hydrogen to form R–H

R–OH
$R–CO_2H$
H–X

→ R–H

- R–MgX
- R–Li
- R_2CuLi

Reaction with acidic deuterium to form R–D

R–OD
$R–CO_2D$
D–X

→ R–D

- R–MgX
- R–Li
- R_2CuLi

Reagents for Making Alkyl/ Acyl Halides

Alcohol to alkyl chloride

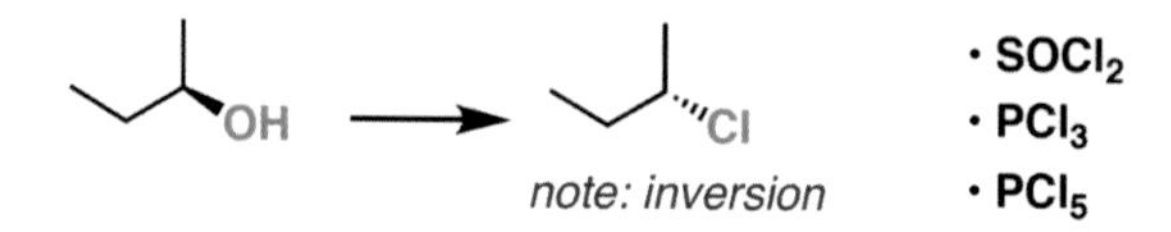

- $SOCl_2$
- PCl_3
- PCl_5

Alcohol to alkyl bromide

OH → Br

note: inversion

- PBr_3
- $SOBr_2$

Alcohol to alkyl sulfonate

OH → O–S(=O)(=O)–R

note: retention

- TsCl
- MsCl
- BsCl

Carboxylic acid to acyl chloride

O, OH → O, Cl

- $SOCl_2$
- PCl_3
- PCl_5

Carboxylic acid to acyl bromide

O, OH → O, Br

- $SOBr_2$
- PBr_3

Reagents Involving Aromatic Rings

Chlorination

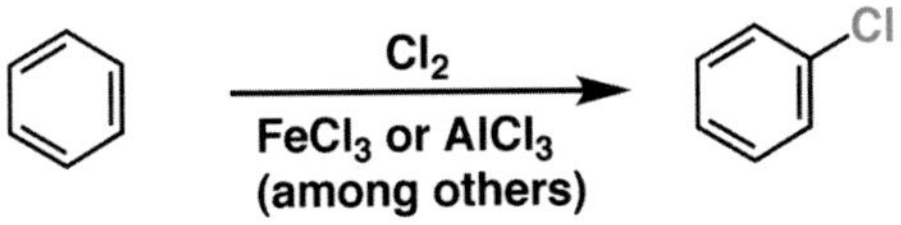

Cl_2; $FeCl_3$ or $AlCl_3$ (among others)

Cl

Bromination

Br_2; $FeBr_3$ or $AlBr_3$

Br

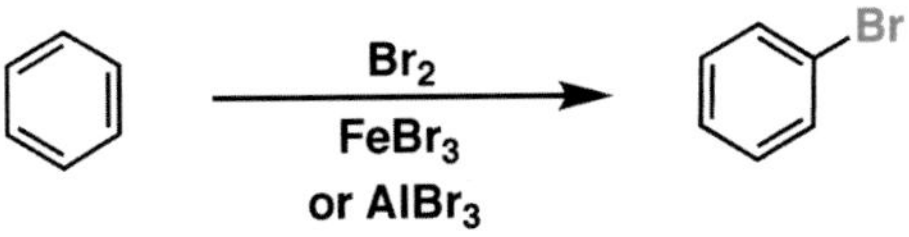

Friedel-Crafts alkylation

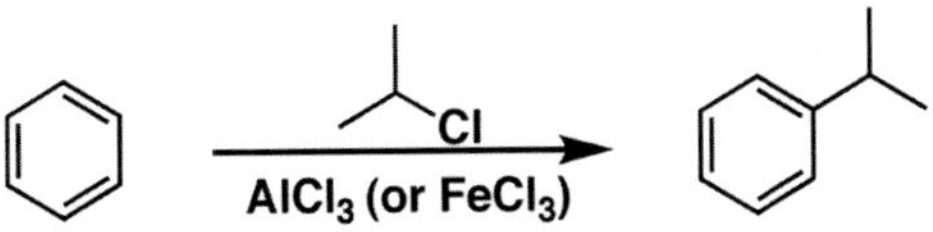

Cl

$AlCl_3$ (or $FeCl_3$)

Works best for secondary and tertiary halides - primary alkyl halides will rearrange

Friedel-Crafts acylation

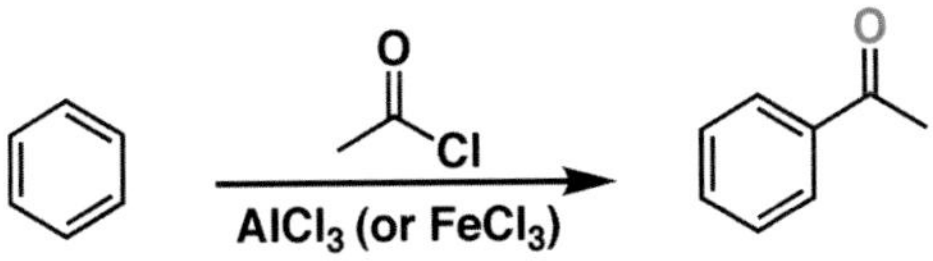

O

Cl

$AlCl_3$ (or $FeCl_3$)

O

Nitration

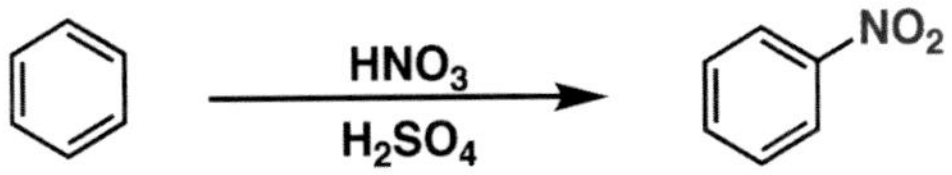

HNO_3; H_2SO_4

NO_2

Sulfonylation

SO_3; H_2SO_4

O S OH O

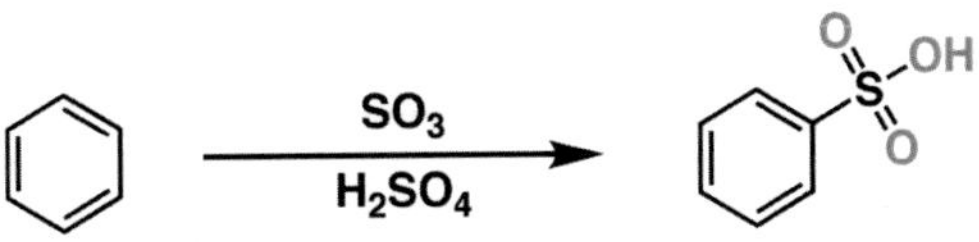

Reduction of nitro group

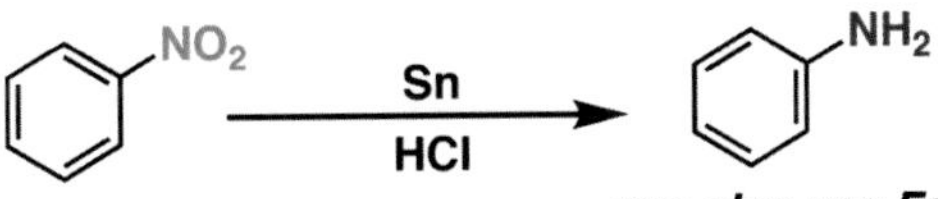

NO_2

Sn; HCl

NH_2

can also use Fe or Zn or Pd/C + H_2

Conversion of aromatic amines to diazonium salts

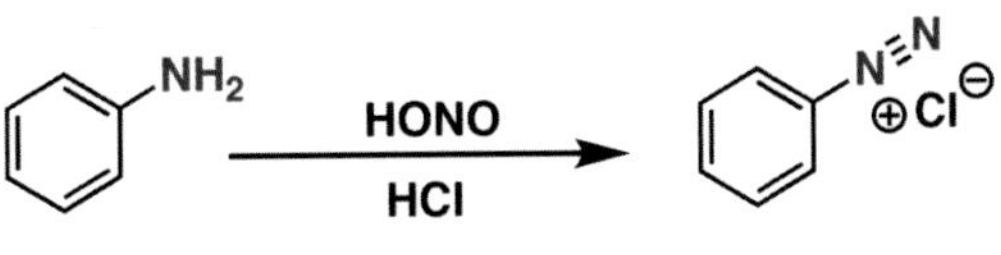

NH_2

HONO; HCl

N≡N ⊕ $Cl^{\ominus}$

Reagents Involving Aromatic Rings

Chlorination of diazonium salts — $PhN_2^+ X^-$ → (CuCl) → $PhCl$

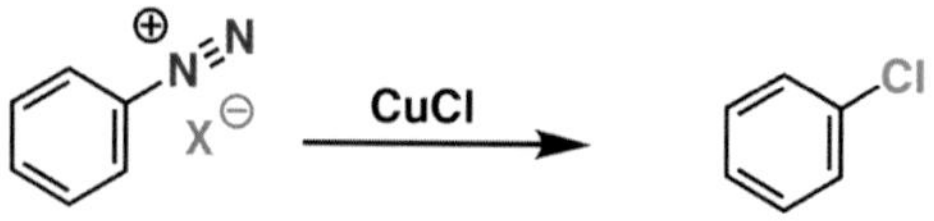

Bromination of diazonium salts — $PhN_2^+ X^-$ → (CuBr) → $PhBr$

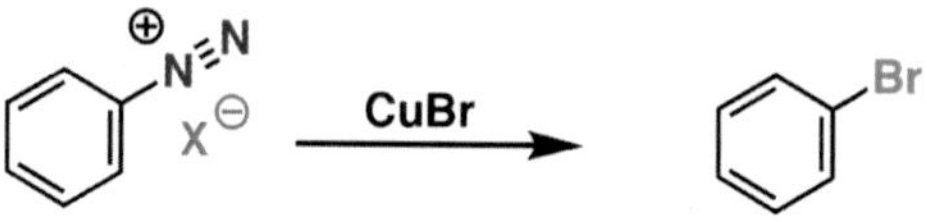

Cyanation of diazonium salts — $PhN_2^+ X^-$ → (CuCN) → $PhCN$

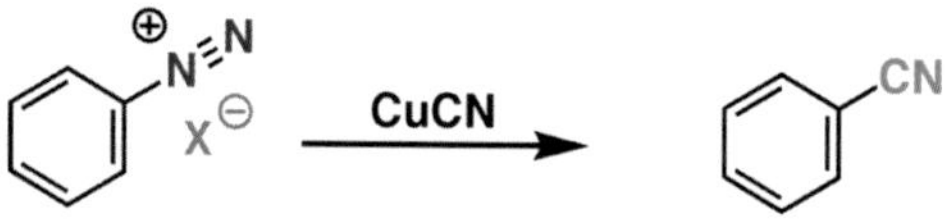

Formation of phenols — $PhN_2^+ X^-$ → (H_2O, heat) → $PhOH$

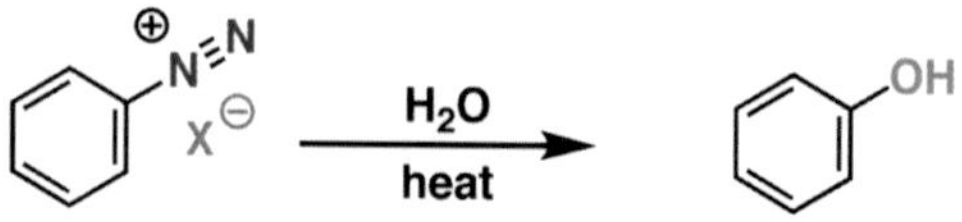

Formation of aryl iodides — $PhN_2^+ X^-$ → (KI, heat) → PhI

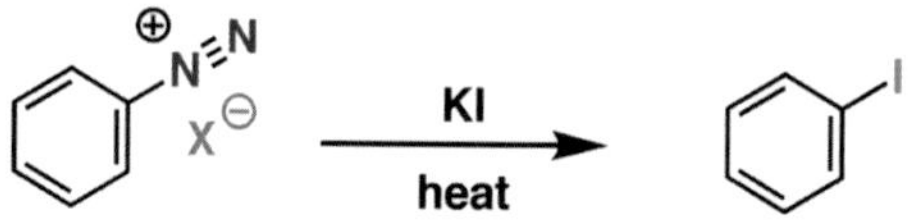

Oxidation of aromatic side chains — $KMnO_4$, heat

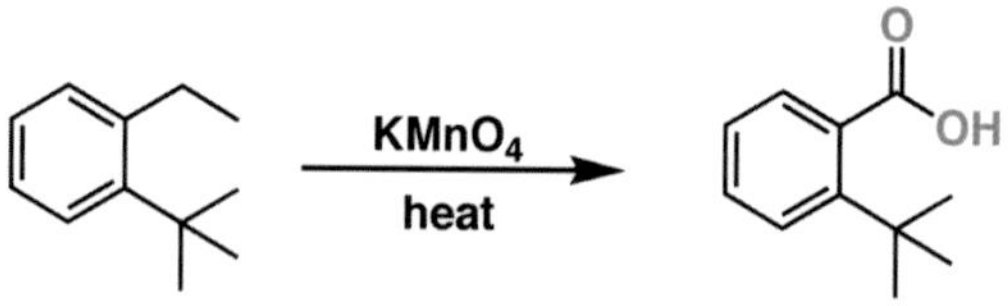

Note that only carbons adjacent to the ring are oxidized, and they must have at least one C-H bond

Hydrogenation of aromatic rings — Pt, Pd, or Ni; H_2 (high pressure), heat

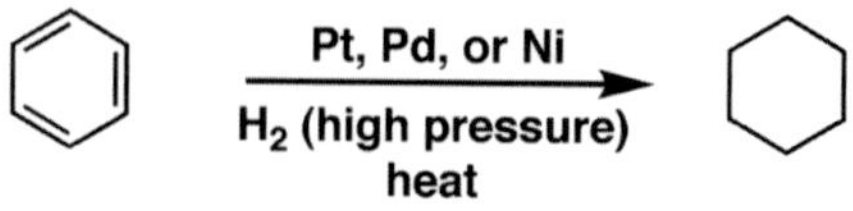

Types of Arrows

1. Reaction Arrow

"this goes to this"

2. Equilibrium Arrows

"reaction goes reversibly between products and reactants"

"reaction goes reversibly, but favors products"

"reaction goes reversibly, but favors reactants".

3. Resonance Arrow

"these two molecules are resonance structures"

4. Curved arrow (double)

"take a pair of electrons from here, and move them to there".

5. Curved arrow (single)

"take a single electron from here, and move it there".

6. Dashed arrow

"we'd like to do this, but haven't done it"

7. Broken arrows

"this doesn't work"

8. Retrosynthesis arrow

"Make this from this"

Types of Solvents

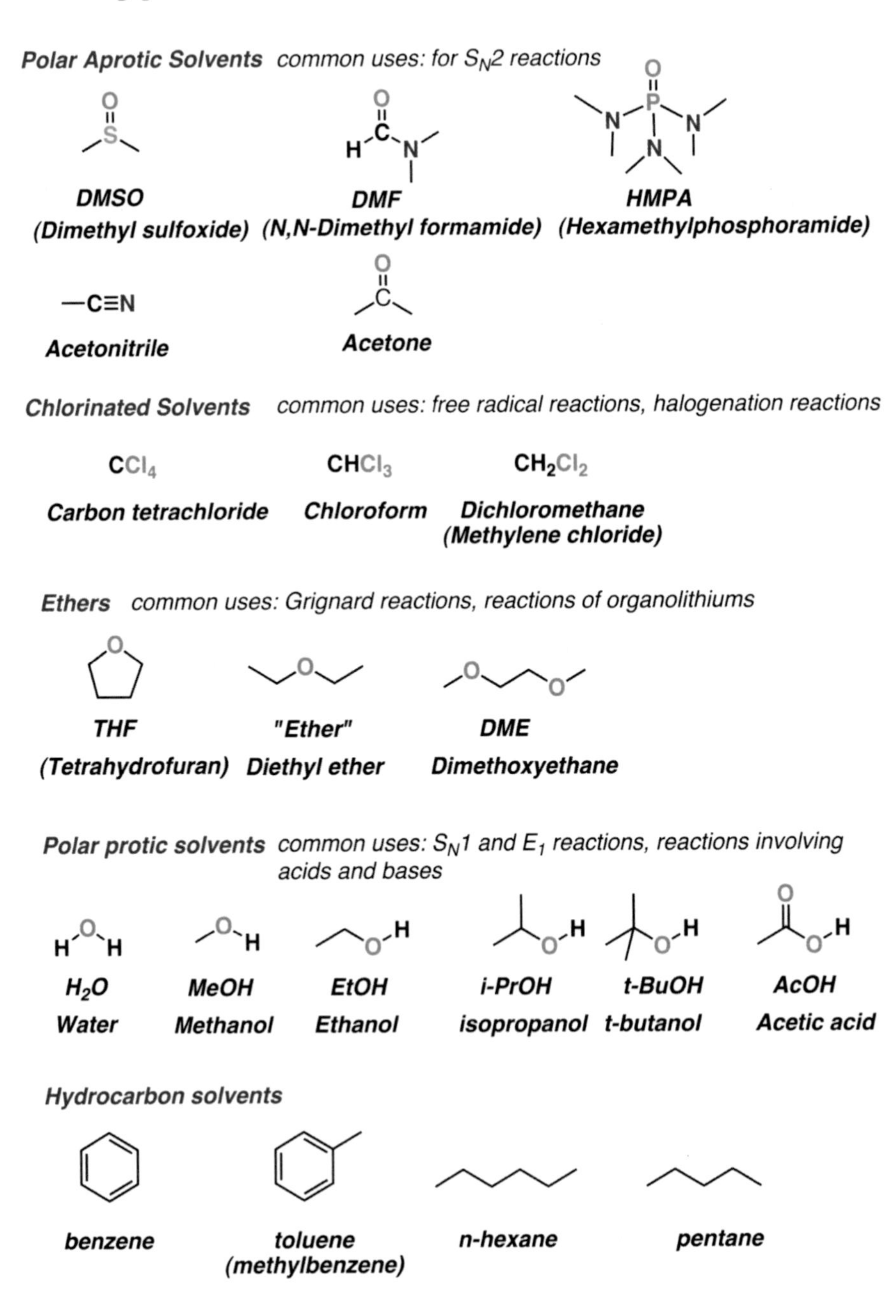

Protecting Groups

Protection of alcohols

R–CH₂–OH —(Me_3SiCl)→ R–CH₂–O–SiMe₃

Silyl ether

Deprotection: TBAF or aqueous acid

R–CH₂–OH —(dihydropyran; **Acid catalyst (e.g. TsOH)**)→ R–CH₂–O–THP

Acetal

Deprotection: aqueous acid

R–CH₂–OH —(acetic anhydride, **(Ac_2O)**)→ R–CH₂–O–C(=O)CH₃

Ester

Deprotection: aqueous acid or base

Protection of aldehydes and ketones

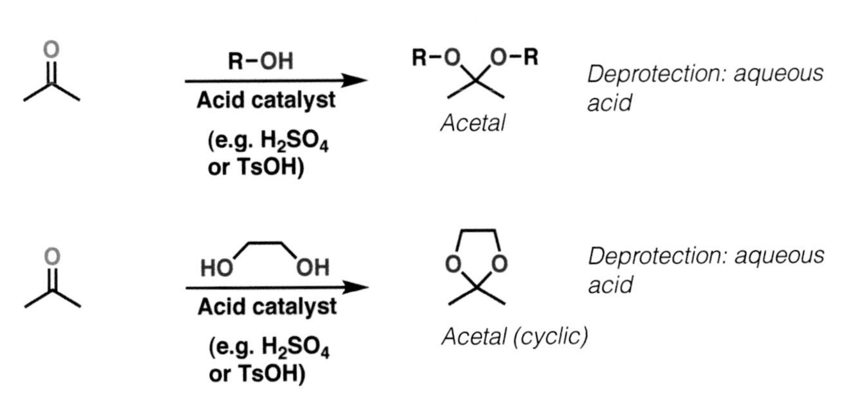

Made in the USA
Middletown, DE
18 July 2019